This book is dedicated to my father-in-law, Dilip Kumar Mukherjee, one of my greatest champions, who left us in December 2011 and did not live to see this work printed. His spirit lives on through my wife Lali and my son Ishan, who give me their continuous love and support.

—*Indranil*

## ABOUT THE AUTHOR

**Dr. Indranil Goswami** has a Ph.D. in Civil Engineering from Johns Hopkins University in Baltimore, Maryland. During the past 20 years he has taught and conducted research at Morgan State University in Baltimore. Dr. Goswami is a registered Professional Engineer in the State of Maryland and he has served as President of the Baltimore Chapter of the Maryland Society of Professional Engineers. He taught PE (civil) and FE review courses in Baltimore from 2001 to 2010. In 1998, Dr. Goswami was nominated for the Bliss Medal for Academic Excellence by the Society of American Military Engineers, and in 2003 he received the Educator of the Year Award from the Maryland Society of Professional Engineers.

# CONTENTS

# PREFACE

This book contains 280 questions divided into seven 40-question clusters. There are two (Chapters 1 and 2) breadth (AM) exams representative of the NCEES guidelines for the Principles and Practice of Civil Engineering. The detailed solutions for these two exams are in Chapters 8 and 9, respectively. These exams are followed by five depth (PM) exams—Chapter 3 (Structural Depth), Chapter 4 (Geotechnical Depth), Chapter 5 (Water Resources & Environmental Depth), Chapter 6 (Transportation Depth), and Chapter 7 (Construction Depth). The detailed solutions for these exams are in Chapters 10 to 14, respectively.

There are two ways you can use this book. Early in your review process, you can use the entire set of 280 questions as a repository of practice problems. The questions have been formatted to be similar to those on the actual exam given by the NCEES. As with any other set of solved problems, remember that reading through *example solutions created by someone else is not adequate preparation for an exam such as this*. I find that when I am reading the solution to an example, unless I police myself consciously, I sometimes don't think about *why the solution starts the way it does*. I find myself checking the arithmetic and that the numbers in step 1 lead to the numbers in step 2 and so on, but I don't ask myself, "Would I have started the solution the same way, based on the specifics of the question?" In any kind of problem solving, the initial setup of the solution is more than half the battle; the rest is usually just arithmetic and algebra.

A second way to utilize the book is to combine one of the breadth (AM) exams with the depth exam of your choice to create a full 8-hour exam. Take this timed exam toward the end of your review, when you feel you are adequately prepared.

The one thing that may be perceived as missing from these exams is qualitative and "look-up" questions. These could literally be from anywhere and how well you are able to answer them depends on (1) your familiarity with these codes and standards and (2) your ability to "see the big picture" in a particular subject. This ability is what I like to call the *difference between knowledge and wisdom*. In order to develop this wisdom, you *must* supplement problem solving with readings from a few well-written and well-organized books. You probably need this level of wisdom only for your depth area, so allocate your limited time accordingly.

In using these and any other practice exams, you should develop and implement test-taking strategies such as making multiple passes through the exam without getting bogged down in unfamiliar areas, practicing effective time management, and effective use of tabbing all materials.

It is perhaps unavoidable that on an exam such as the PE, you will spend some time on a particular question before realizing that it is "not going well" and decide to move on to the next one. If you have prepared well, you should have a very good handle on your strengths and weaknesses. This will allow you to quickly recognize the intent of a question and make a quick decision on whether to do the problem right away, or whether to come back to it in the next pass. The quicker you make this decision, the better off you are in terms of effective use of time. Make sure you have some personal method to mark eliminated (or favored) choices before you move on, so that if you come back to this problem and are forced to make a guess, you make a higher probability guess.

I hope you find this book useful and I wish you good luck on the PE exam.

*Indranil Goswami*

# ACKNOWLEDGMENTS

I would like to acknowledge several people without whose input this book could not have been written in its present form. As it stands, even after rigorous proofreading, I am sure there are instances of ambiguity or lack of elegance in the wording of some of the questions, but my primary purpose in writing this book has been to present questions in the formatting likely to be on the actual exam given by the NCEES. In content, I have tried to remain true to the official syllabus for the exam, as it stands today. I would sincerely appreciate reader feedback on these questions, so that they may be improved in later editions.

There are many who I should thank for their valuable input in the development of these questions. First and foremost, many thanks go out to Ravindra Koirala, who provided a lot of suggestions, particularly about the structural problems. Valuable suggestions have also come from those who have attended my review course. Without this input, many of the problems would have inconsistencies and ambiguities that would make them less valuable to the reader.

I would also like to thank Larry Hager at McGraw-Hill for his editorial leadership of this project. His constant accessibility has always reassured me that I have someone to approach with any ideas.

Thanks are also due to Manisha Singh and her staff at Cenveo for managing the editing, proofreading, and design of the final document.

# Civil Engineering PE Practice Exams

## Breadth and Depth

# 1

# Breadth Exam No. 1

The following set of questions numbered 1 to 40 is representative of a 4-hour breadth exam according to the syllabus and guidelines for the Principles and Practice (P&P) of Civil Engineering Examination administered by the National Council of Examiners for Engineering and Surveying (NCEES). Detailed solutions are on page 177.

**I**

Water flows in a rectangular open channel at a normal depth of 4.6 ft as shown. Assume Manning's roughness coefficient (constant with depth) = 0.014. The longitudinal slope of the channel bed is 0.4%. The flow rate (ft³/sec) is most nearly:

A.  600
B.  700
C.  800
D.  900

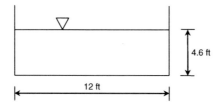

4.6 ft

12 ft

**2**

Water flows by gravity from reservoir A (surface elevation 340 ft) to reservoir B (surface elevation 270 ft) through a system of cast-iron pipes. The characteristics of the pipe system are given below:

Length = 2,500 ft
Diameter = 24 in.
Friction factor = 0.02
Minor loss equivalent length = 55 ft

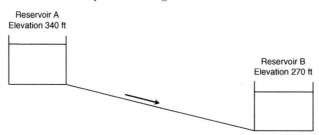

Reservoir A
Elevation 340 ft

Reservoir B
Elevation 270 ft

The flow rate (ft³/sec) is most nearly:

A.  42
B.  55
C.  65
D.  76

**3**

The 1-hr unit hydrograph of excess precipitation is described by the following data:

| Time (hr) | 0.0 | 1.0 | 2.0 | 3.0 | 4.0 | 5.0 |
|---|---|---|---|---|---|---|
| Discharge Q (ft³/sec/in.) | 0 | 30 | 95 | 125 | 50 | 0 |

A storm produces the following pattern of excess precipitation – 1.7 in. of excess precipitation during the first hour, followed by 0.8 in. during the second hour. The stream discharge (ft³/sec) at the end of the second hour is most nearly:

A.  140
B.  155
C.  170
D.  185

**4**

A flow rate = 20 ft³/sec flows through a 30-in.-diameter concrete pipe (Manning's $n$ constant with depth = 0.013). The following data is given:

Pipe length = 800 ft
Pipe invert elevation at upstream end of pipe = 275.64 ft
Pipe invert elevation at downstream end of pipe = 270.96 ft

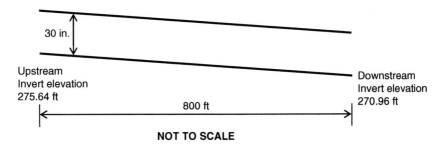

The depth of flow (in.) in the pipe is most nearly:

A.  14.0
B.  15.3
C.  17.5
D.  21.8

## 5

A cantilever reinforced-concrete retaining wall is shown. The friction angle between the wall footing and the underlying soil is 20°.

Elevation of the top of the backfill = 257.00 ft
Elevation of the top of the footing = 243.00 ft
Elevation of the bottom of the footing = 240.00 ft

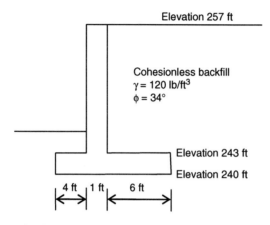

Elevation 257 ft

Cohesionless backfill
$\gamma = 120$ lb/ft$^3$
$\phi = 34°$

Elevation 243 ft

Elevation 240 ft

4 ft   1 ft   6 ft

The factor of safety for sliding of the wall is most nearly:

A. 1.27
B. 0.93
C. 1.45
D. 1.78

## 6

A soil sample weighs 3.64 lb and has volume = 0.031 ft$^3$. Specific gravity of soil solids = 2.65. Water is added to the soil until it bleeds. The added water has volume = 5.6 fl. oz. The dry unit weight of the original soil sample (lb/ft$^3$) is most nearly:

A. 101
B. 104
C. 107
D. 110

## 7

A clay soil deposit has the following characteristics:

Unit weight = 122 lb/ft³
Angle of internal friction = 10°
Unconfined compression strength = 2,400 lb/ft²

A square footing is to be used to support a column load = 80 k. Bottom of footing is 3 ft below the ground surface. Minimum factor of safety for ultimate bearing capacity is 3.0. The minimum footing size required (ft) is most nearly:

A. 6
B. 5
C. 4
D. 3

## 8

An anaerobic digester has volume = 25,000 gal. The influent sludge flow rate is 4 MGD. Influent flow contains TSS = 1,100 mg/L. Suspended solids are 70% volatile. The volatile solids load (lb-VSS/ft³-day) is most nearly:

A. 6.9
B. 7.2
C. 7.4
D. 7.7

## 9

A simply supported steel beam (span = 25 ft) supports a uniformly distributed load = 2.75 k/ft. The allowable bending stress for the chosen grade of steel = 32 ksi. The lightest W-section (choose from the table below) is:

A. W12 × 50
B. W12 × 53
C. W12 × 58
D. W12 × 65

(*Continued on next page*)

| Section | $A_g$ (in.$^2$) | $d$ (in.) | $S_x$ (in.$^3$) | $S_y$ (in.$^3$) |
|---------|---------|-----------|-----------|-----------|
| W12 × 79 | 23.2 | 12.4 | 107.0 | 35.8 |
| W12 × 72 | 21.1 | 12.3 | 97.4 | 32.4 |
| W12 × 65 | 19.1 | 12.1 | 87.9 | 29.1 |
| W12 × 58 | 17.0 | 12.2 | 78.0 | 21.4 |
| W12 × 53 | 15.6 | 12.1 | 70.6 | 19.2 |
| W12 × 50 | 14.6 | 12.2 | 64.2 | 13.9 |

## 10

A beam is shown below. Which of the given diagrams represents a valid shape for the bending moment diagram of the beam?

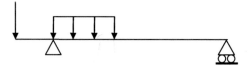

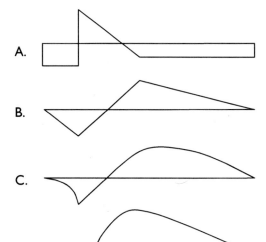

A.

B.

C.

D.

## 11

A steel beam has a singly symmetric I-shaped cross section as shown. The elastic section modulus (in.$^3$) about the major centroidal axis is most nearly:

A. 54
B. 63
C. 66
D. 73

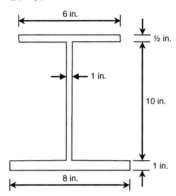

## 12

A parabolic vertical curve is being designed to connect two grades with $G_1 = +5\%$ and $G_2 = -3\%$. If PVC is at station $117 + 50.00$, the elevation (ft) of the PVT is most nearly:

A. 307.21
B. 309.45
C. 311.12
D. 313.02

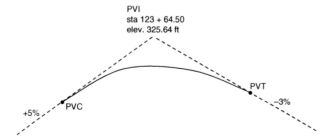

## 13

A pavement-based dual loop detector is used to monitor speeds of a stream of vehicles in one travel lane. The following speeds (mph) were recorded:

45, 51, 58, 47, 50, 53, 51, 42, 61, 51

The traffic flow rate during this period is estimated to be 1,450 vphpl. The mean headway (sec) is most nearly:

A. 1.9
B. 2.1
C. 2.3
D. 2.5

## 14

For a municipality in New York State, the recommended side friction coefficient under icy conditions is 0.08. For a ramp with design speed = 40 mph and 4% superelevation, the minimum curve radius (ft) is most nearly:

A. 1,000
B. 880
C. 1,050
D. 1,200

## 15

A highway with a straight alignment slopes down to a stop line as shown. The design speed of the highway is 45 mph. The stopping sight distance (ft) is most nearly:

A. 395
B. 360
C. 230
D. 195

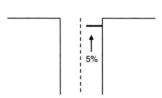

## 16

A horizontal circular curve has PC at coordinates (ft) 4123.64 N, 1064.32 W.
Curve radius = 1,030 ft. The tangent at the PC has bearing S42°30'W. The length
of curve = 646.35 ft. The coordinates of the PI are:

A.  3877.23 N, 1290.12 W
B.  3897.84 N, 1310.73 W
C.  4370.05 N, 838.52 W
D.  4349.44 N, 817.91 W

## 17

Water flows through a circular conduit of diameter 42 in. At what depth of flow (in.)
does the pipe convey maximum flow?

A.  21
B.  30
C.  40
D.  42

## 18

BOD tests were conducted on a water sample. Incubation was for 5 days at 20°C.
The following data are given:

Sample volume = 15 mL
Volume of BOD bottle = 300 mL
At $t = 0$, dissolved oxygen of the diluted sample = 6.3 mg/L
At $t = 5$ days, dissolved oxygen of the diluted sample = 2.7 mg/L
Deoxygenation rate constant (base-10, 20°C) = 0.1 $day^{-1}$

The ultimate BOD of the water sample (mg/L) is most nearly:

A.  95
B.  105
C.  115
D.  125

## 19

Grain size distribution for a soil sample is shown in the curve below. Liquid limit = 34%. Plastic limit = 19%. What is the USCS classification?

A.  SW
B.  SP
C.  ML
D.  CL

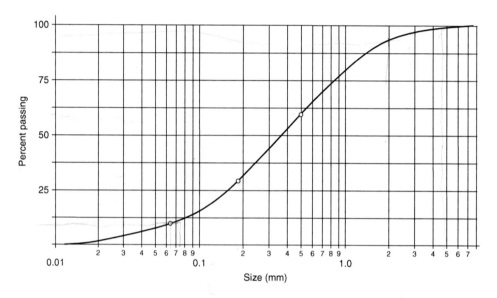

## 20

For the soil described above, the AASHTO soil classification is:

A.  A-6
B.  A-2-6
C.  A-2-7
D.  A-7

## 21

A three-hinged arch (internal hinge at B) is shown. The arch supports a bridge deck whose design load is 9 k/ft (uniformly distributed on horizontal projection). The horizontal thrust reaction (k) at A is most nearly:

A. 124 (to the left)
B. 124 (to the right)
C. 168 (to the left)
D. 168 (to the right)

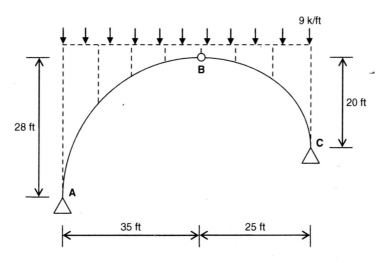

## 22

A concrete mix contains cement, moist sand, and moist coarse aggregate in the following proportions by weight: 1:1.8:2.6. The following specifications are given:

| | |
|---|---|
| Cement | specific gravity = 3.15 |
| SSD sand (m.c. = 0.5%) | specific gravity = 2.70 |
| SSD coarse aggregate (m.c. = 0.7%) | specific gravity = 2.60 |
| Added water | 5.8 gal per sack cement |
| Air | 3% (by volume) |

(*Continued on next page*)

The aggregates used for mixing the concrete had the following properties:

Wet sand: moisture content = 6%

Wet coarse aggregate: moisture content = 4%

The water content (gal/sack) of the concrete is most nearly:

A. 6.1

B. 7.1

C. 7.9

D. 8.5

## 23

A crane is used to lift a load of 16 tons as shown. The crane cabin has a ballasted weight of 24 tons. The allowable soil pressure is 2,800 lb/ft². If the crane is supported by 4 outriggers, as shown, the minimum contact area (ft²) of the outrigger pads is most nearly:

A. 9.0

B. 12.3

C. 16.0

D. 20.3

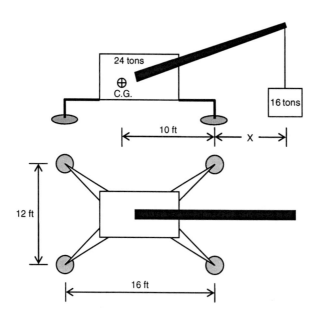

## 24

A project consists of 10 activities as outlined in the table below. All relationships are finish-to-start unless otherwise noted.

| Activity | Duration (months) | Predecessors | Successors |
|----------|-------------------|--------------|------------|
| A | 5 | — | C, D, E |
| B | 3 | — | H |
| C | 3 | A | F |
| D | 4 | A | G |
| E | 3 | A | H |
| F | 5 | C | J |
| G | 4 | D (FF LAG = 5) | J |
| H | 3 | B, E | I |
| I | 2 | H | J |
| J | 2 | F, G, I | — |

The activity on arrow representation of the project is also shown below.

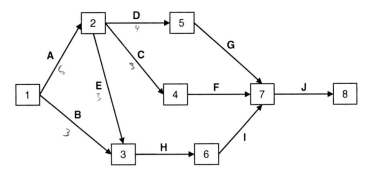

The critical path for the project is

A. ACFJ

B. ADGJ

C. BHIJ

D. AEHIJ

## 25

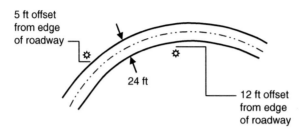

5 ft offset from edge of roadway

24 ft

12 ft offset from edge of roadway

The centerline of the local horizontal alignment of a two-lane highway is a circular curve with radius = 915 ft. The length of curve is 600 ft. Trees are located adjacent to the highway, as shown. Assume standard lane width = 12 ft. The safe design speed (mph) based on adequate sight distance is most nearly:

A. 45
B. 50
C. 55
D. 60

## 26

A gravity dam has a cutoff wall at the upstream end as shown in the figure below. The width of the base of the dam is 100 ft and the length of the dam is 150 ft. An impermeable layer exists approximately 50 ft below the bottom of the reservoir.

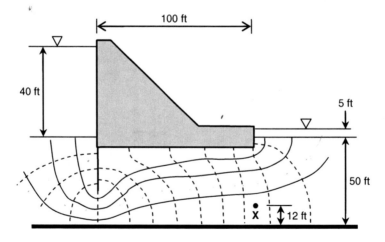

100 ft

40 ft

5 ft

50 ft

X   12 ft

The upstream reservoir depth is 40 ft and the tailwater depth is 5 ft. The hydraulic conductivity of the soil in the 50-ft-thick soil layer underlying the dam is 0.0025 ft/sec. The pressure (psig) at point X (located 12 ft above the top of the impermeable layer) is most nearly:

A.  45
B.  37
C.  30
D.  22

## 27

A hoisting mechanism uses a cable system as shown below. If the load W = 3 tons, the effort required (lb) at point B is most nearly:

A.  1345
B.  1715
C.  2065
D.  2350

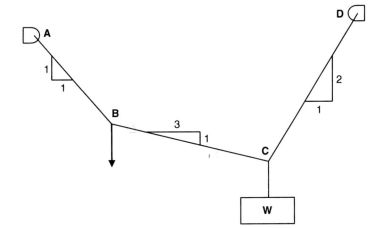

## 28

A square footing is embedded 3 ft in a sand layer as shown. The concentric column load is 150 k. A water table is located 10 ft below the bottom of footing. Minimum factor of safety for ultimate bearing capacity = 3.0. Unit weight of soil = 125 lb/ft³. Unit weight of concrete = 145 lb/ft³.

Angle of internal friction of sand = 32°

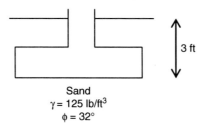

3 ft

Sand
$\gamma = 125$ lb/ft³
$\phi = 32°$

The minimum footing size required (ft) is most nearly:
A.  4.5
B.  5.0
C.  5.5
D.  6.0

## 29

Construction of a 50-ft-wide roadway requires fill between station 10 + 0.00 and station 11 + 0.00. The table below shows required depth of fill (ft) at various horizontal offsets (ft) from the centerline of the roadway alignment. Earthwork is to be calculated by the average end area method. The volume of fill (yd³) is most nearly:
A.  2,600
B.  3,500
C.  7,000
D.  14,000

| Sta. 10 + 00 | F 22 | F 14 | F 10 | F 6 | F 4 |
| --- | --- | --- | --- | --- | --- |
| | −50 | −25 | 0.0 | +25 | +31 |
| Sta. 11 + 00 | F 20 | F 15 | F 9 | F 7 | F 5 |
| | −47 | −25 | 0.0 | +25 | +33 |

**30**

An excavation trench is 5 ft deep, 6 ft wide, and 95 ft long. According to OSHA standards, how many ladders are required to provide egress from the trench?

A. 4
B. 3
C. 2
D. 1

 **31**

An excavator has a bucket capacity of 2.8 yd$^3$. Its operation cycle consists of the following phases: (a) excavation time = 45 sec, (b) travel time (two-way) = 4 min, and (c) dumping/transfer time = 30 sec. Assume an overall efficiency factor for the excavator = 85%. The quantity of excavated material = 50,000 ft$^3$ (bank measure). The material has the following properties:

Swell = 20%
Unit weight = 120 lb/ft$^3$
Water content = 30%

The number of days required to complete the job, assuming 8-hr workdays, is most nearly:

A. 10
B. 11
C. 12
D. 13

## 32

For the truss shown below, the member force (k) in member DI is most nearly:

A.  22 (compression)
B.  22 (tension)
C.  39 (compression)
D.  39 (tension)

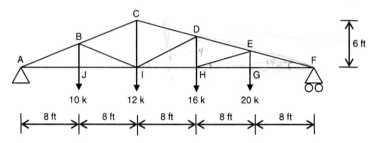

## 33

A parabolic vertical curve joins a grade of −4% to a grade of +6%. The PVI is at station 10 + 56.30. Elevation of the PVI is 432.65 ft. The curve passes under a bridge structure at station 12 + 00.00. The elevation of the bottom of the bridge girders is 470.00 ft. The minimum vertical clearance under the bridge is 14 ft 6 in. The required length of the curve (ft) is most nearly:

A.  1,665
B.  1,865
C.  2,065
D.  2,265

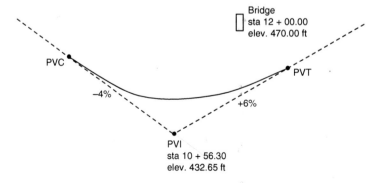

## 34

A local municipality must make a determination about the feasibility of repairing a bridge. The estimated useful life of the bridge is 25 years. The MARR is 8%. The relevant cost estimates are shown below. All costs are shown in thousands of dollars.

|  | Existing | Repair |
|---|---|---|
| Initial Cost | 0 | 75 |
| Annual Cost (maintenance) Years 1–15 Years 16–25 | 10 14 | 3 4 |
| Salvage Value | 5 | 20 |

The benefit-cost ratio of making repairs is most nearly:

A.  0.90

B.  0.95

C.  1.05

D.  1.10

## 35

A room is 35 ft × 25 ft in plan. Ceiling height is 14 ft. Openings for doors and windows total 85 ft². The following data is given for plastering and painting operations.

Plaster and paint crew:

    1 supervisor $30/hr
    1 laborer $12/hr
    2 painter $18/hr

Plastering productivity = 50 ft²/L.H.

Painting productivity = 150 ft²/L.H.

The estimated labor cost for plastering and painting the room (walls and ceiling) is most nearly:

A.  $1,020

B.  $1,280

C.  $1,460

D.  $1,690

## 36

A 4-ft-wide wall footing constructed on the surface of a silty sand soil is shown in Case A. If the footing is embedded to a depth of 30 in. (as shown in case B), the percent increase of the bearing capacity is most nearly:

A. 50
B. 90
C. 140
D. 190

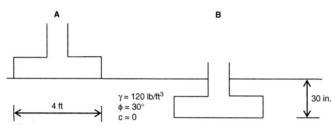

## 37

A 10-ft-thick layer of soft clay (site A) experiences a maximum consolidation settlement of 10 in., with 8 in. of that occurring after 5 years. The soil above the clay layer is silty sand and the layer underneath it is sandy gravel. In another location (site B), an identical clay layer has a silty sand layer above and a clayey shale below it. If the preconsolidation pressure and the increased pressure on both layers are the same, the settlement (in.) after 5 years at location B is most nearly:

A. 2.7
B. 4.5
C. 6.2
D. 7.3

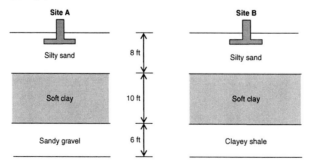

## 38

A landfill accepts municipal solid waste from town A (population 20,000). The landfill has been in operation for 6 years and has remaining capacity of 1,000,000 yd³. A new subdivision (population 5,000) is developed and populated. The waste generation rate is 5 lb/capita-day. The average compacted density in the landfill is 40 lb/ft³. The reduction in the service life (years) of the landfill due to the addition of the new subdivision is most nearly:

A. 6
B. 8
C. 10
D. 12

##  39

A reinforced-concrete cantilever wall is used to retain an engineered fill (on the right of the stem) as shown. Drains are designed so that water pressure does not build up behind the wall. Which of the following patterns shows the correct placement of flexural reinforcement in the wall?

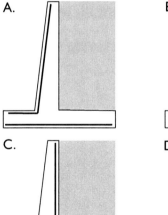
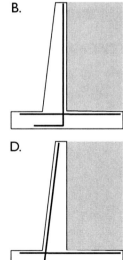

## 40

For the 2D rigid frame shown, the bending moment (k-ft) at joint B is most nearly:

A. 550
B. 450
C. 270
D. 360

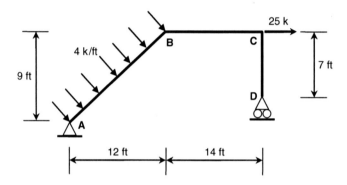

**END OF BREADTH EXAM NO. I**

**USE THE ANSWER SHEET ON THE NEXT PAGE**

## Breadth Exam No. 1: Answer Sheet

| | | | | |
|---|---|---|---|---|
| 1 | Ⓐ | Ⓑ | Ⓒ | Ⓓ |
| 2 | Ⓐ | Ⓑ | Ⓒ | Ⓓ |
| 3 | Ⓐ | Ⓑ | Ⓒ | Ⓓ |
| 4 | Ⓐ | Ⓑ | Ⓒ | Ⓓ |
| 5 | Ⓐ | Ⓑ | Ⓒ | Ⓓ |
| 6 | Ⓐ | Ⓑ | Ⓒ | Ⓓ |
| 7 | Ⓐ | Ⓑ | Ⓒ | Ⓓ |
| 8 | Ⓐ | Ⓑ | Ⓒ | Ⓓ |
| 9 | Ⓐ | Ⓑ | Ⓒ | Ⓓ |
| 10 | Ⓐ | Ⓑ | Ⓒ | Ⓓ |
| 11 | Ⓐ | Ⓑ | Ⓒ | Ⓓ |
| 12 | Ⓐ | Ⓑ | Ⓒ | Ⓓ |
| 13 | Ⓐ | Ⓑ | Ⓒ | Ⓓ |
| 14 | Ⓐ | Ⓑ | Ⓒ | Ⓓ |
| 15 | Ⓐ | Ⓑ | Ⓒ | Ⓓ |
| 16 | Ⓐ | Ⓑ | Ⓒ | Ⓓ |
| 17 | Ⓐ | Ⓑ | Ⓒ | Ⓓ |
| 18 | Ⓐ | Ⓑ | Ⓒ | Ⓓ |
| 19 | Ⓐ | Ⓑ | Ⓒ | Ⓓ |
| 20 | Ⓐ | Ⓑ | Ⓒ | Ⓓ |

| | | | | |
|---|---|---|---|---|
| 21 | Ⓐ | Ⓑ | Ⓒ | Ⓓ |
| 22 | Ⓐ | Ⓑ | Ⓒ | Ⓓ |
| 23 | Ⓐ | Ⓑ | Ⓒ | Ⓓ |
| 24 | Ⓐ | Ⓑ | Ⓒ | Ⓓ |
| 25 | Ⓐ | Ⓑ | Ⓒ | Ⓓ |
| 26 | Ⓐ | Ⓑ | Ⓒ | Ⓓ |
| 27 | Ⓐ | Ⓑ | Ⓒ | Ⓓ |
| 28 | Ⓐ | Ⓑ | Ⓒ | Ⓓ |
| 29 | Ⓐ | Ⓑ | Ⓒ | Ⓓ |
| 30 | Ⓐ | Ⓑ | Ⓒ | Ⓓ |
| 31 | Ⓐ | Ⓑ | Ⓒ | Ⓓ |
| 32 | Ⓐ | Ⓑ | Ⓒ | Ⓓ |
| 33 | Ⓐ | Ⓑ | Ⓒ | Ⓓ |
| 34 | Ⓐ | Ⓑ | Ⓒ | Ⓓ |
| 35 | Ⓐ | Ⓑ | Ⓒ | Ⓓ |
| 36 | Ⓐ | Ⓑ | Ⓒ | Ⓓ |
| 37 | Ⓐ | Ⓑ | Ⓒ | Ⓓ |
| 38 | Ⓐ | Ⓑ | Ⓒ | Ⓓ |
| 39 | Ⓐ | Ⓑ | Ⓒ | Ⓓ |
| 40 | Ⓐ | Ⓑ | Ⓒ | Ⓓ |

# 2

# Breadth Exam No. 2

The following set of questions numbered 101 to 140 is representative of a 4-hour breadth exam according to the syllabus and guidelines for the Principles and Practice (P&P) of Civil Engineering Examination administered by the National Council of Examiners for Engineering and Surveying (NCEES). Detailed solutions are on page 199.

## 101

For the plane truss shown below, the axial force (k) in member AH is most nearly:

A.  89 (compression)
B.  89 (tension)
C.  79 (tension)
D.  79 (compression)

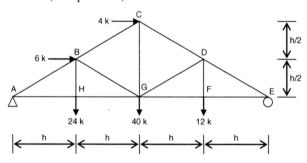

## 102

A temporary warning sign is constructed at a worksite by using a 14-in.-diameter bucket filled with ballast so as to serve as a counterweight (see figure). If the maximum (3-sec gust) wind pressure is 55 psf (ignore the wind pressure on the bucket), the minimum required weight (lb) of the counterweight is most nearly:

A.  165
B.  280
C.  330
D.  560

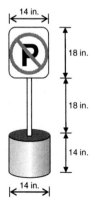

## 103

A steel angle is anchored to a wall as shown. Longitudinal spacing between anchor bolts is 18 in. The heel of the angle is continuously supported by timber blocking (height = 1 in., thickness = 0.5 in.). The angle supports a wall panel centered at a horizontal distance of 4 in. from the heel of the angle. The axial tension in the anchor bolts (k) is most nearly:

A. 65
B. 100
C. 130
D. 200

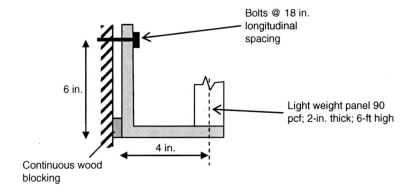

## 104

At a particular point on a beam that is subject to transverse loads, the following stresses have been measured. The x-axis is oriented along the longitudinal axis of the beam.

Bending stress:    $\sigma_{xx} = 28.5$ ksi (tension)

$\sigma_{yy} = 0$

$\tau_{xy} = 14.8$ ksi

The steel used for the beam has the following ultimate stresses:

$\sigma_{ult} = 50$ ksi (tension)

$\tau_{ult} = 30$ ksi

(*Continued on next page*)

The factor of safety based on stress is most nearly:

A. 1.45
B. 1.75
C. 1.95
D. 2.05

## 105

A 3-span continuous bridge is shown below. The design live load is a uniformly distributed load of 800 lb/ft in each design lane. For design of the pier at C, what is the critical live load placement?

A. A
B. B
C. C
D. D

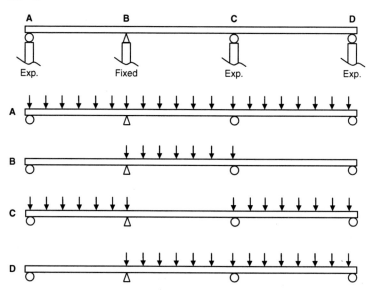

## 106

A residential structure has a floor supported by a system of joists spaced every 36 in. as shown below. The combined floor load is 90 psf. The allowable bending stress in the timber joists is 1,700 psi. The required section modulus (in.³) of the timber joists is most nearly:

A. 30
B. 50
C. 80
D. 120

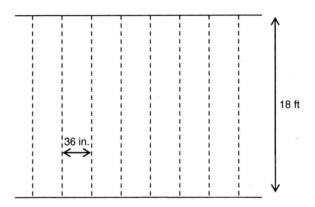

## 107

For the cable system shown, the tension (k) in cable AB is most nearly:
A. 16
B. 19
C. 22
D. 25

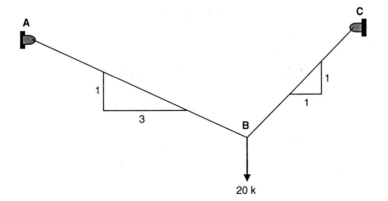

## 108

The tower truss shown below has frictionless hinges at each node. A vertical load acts at node A. The zero-force members are:

A. HI, HG, BD, CD, EF
B. HI, BD, CD, AD, EF
C. HI, HG, DE, FG, EF
D. AD, EF, HI, BD, CD

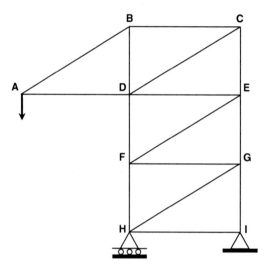

## 109

Results from sieve analysis of a soil sample have been summarized in the table below. Atterberg tests resulted in: liquid limit = 54; plastic limit = 23. What is the USCS classification for the soil?

A. GP
B. GP-GC
C. GP-GM
D. GW-GC

| Sieve Size | % Passing |
|---|---|
| 2 in. | 95 |
| 1 in. | 85 |
| ½ in. | 60 |
| No. 4 | 41 |
| No. 10 | 31 |
| No. 40 | 22 |
| No. 200 | 10 |

## 110

Results from sieve analysis of a soil sample have been summarized in the following table. Atterberg tests resulted in: liquid limit = 54; plastic limit = 23. What is the AASHTO classification for the soil?

A.  A-7
B.  A-2-7
C.  A-6
D.  A-2-6

| Sieve Size | % Passing |
|---|---|
| 2 in. | 95 |
| 1 in. | 85 |
| ½ in. | 60 |
| No. 4 | 41 |
| No. 10 | 31 |
| No. 40 | 22 |
| No. 200 | 10 |

## III

A cantilever retaining wall is shown below. The resultant active earth pressure per unit length (lb/ft) of the wall is most nearly:

A. 280

B. 1,900

C. 4,260

D. 8,530

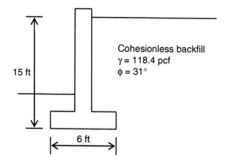

Cohesionless backfill
$\gamma = 118.4$ pcf
$\phi = 31°$

15 ft

6 ft

## II2

A soil sample, taken from a borrow pit, has a specific gravity of soil solids of 2.66. Six samples with varying moisture content were prepared for the standard Proctor test. The results of the standard Proctor test are shown below.

| Sample | Weight of Soil (lb) | Water Content (%) |
|--------|--------------------|-------------------|
| 1 | 3.20 | 12.8 |
| 2 | 3.78 | 13.9 |
| 3 | 4.40 | 15.0 |
| 4 | 4.10 | 15.7 |
| 5 | 3.70 | 16.6 |
| 6 | 3.30 | 18.1 |

The maximum dry unit weight (lb/ft³) is most nearly:

A. 85

B. 90

C. 100

D. 115

# 113

Given the phase diagram for a soil sample, as shown below, the porosity is most nearly:

A.  0.42%

B.  3.3%

C.  9.1%

D.  30%

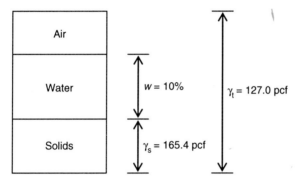

# 114

A clayey sand (SC) was compacted into a 1/30-cu-ft-cylindrical Proctor steel mold with the following results:

    Weight of wet soil: 4.18 lb
    Weight of dry soil: 3.67 lb
    Specific gravity of solids: 2.73

The percent saturation of this compacted sample is most nearly:

A.  70%

B.  50%

C.  15%

D.  85%

## 115

A dam of length 120 ft (perpendicular to plane shown) is constructed of an impermeable soil. The dam overlays a sand bed with a thickness of 6.5 ft, as shown. The permeability of the sand bed is 9.5 ft/day.

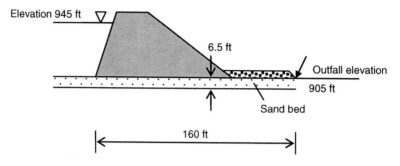

The discharge due to seepage (gal/min) is most nearly:

A. 7.5
B. 9.6
C. 10.5
D. 11.0

## 116

A direct shear test is performed on a soil sample. The sample is capped by porous end-plates and subjected to normal stress in a split-box apparatus as shown in the figure. The sample is then subjected to shear forces that induce shear failure on the interface plane. Results are summarized below:

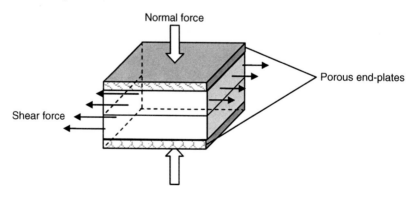

Sample A, tested at a total vertical stress equal to 1,000 psf, failed at a shear stress of 675 psf, while sample B, tested at a total vertical stress of 3,000 psf, failed at a shear stress of 2,025 psf.

Based upon the results of this direct shear test, the shear strength parameters for this soil are most nearly:

A.  $c = 0$ psf and $\phi = 34°$
B.  $c = 200$ psf and $\phi = 34°$
C.  $c = 0$ psf and $\phi = 42.5°$
D.  $c = 200$ psf and $\phi = 42.5°$

# 117

A stream has the following characteristics:

Flow rate = 18 ft$^3$/sec
Temperature = 12°C
BOD$_5$ = 2.0 mg/L
D.O. = 5.1 mg/L

A factory discharges a wastewater stream into the river at point A. The wastewater has the following characteristics:

Flow rate = 750 gal/min
Temperature = 37°C
BOD$_5$ = 9.6 mg/L
D.O. = 1.7 mg/L

The dissolved oxygen (mg/L) of the river-wastewater mix immediately downstream of point A is most nearly:

A.  4.6
B.  4.8
C.  4.9
D.  5.0

## 118

A wastewater treatment plant treats 3 MGD of wastewater using an array of primary clarifiers in parallel configuration. The plant maintenance manual requires that satisfactory operations be sustained with one clarifier taken offline for maintenance. The wastewater has TSS = 180 mg/L. The maximum solids load on each clarifier is 800 lb-TSS/day. The number of clarifier units required is most nearly:

A.  5
B.  6
C.  7
D.  8

## 119

The 1-hr unit hydrograph of excess precipitation is shown in the table below.

| Time (hr)             | 0 | 1  | 2  | 3   | 4  | 5 |
|-----------------------|---|----|----|-----|----|---|
| Discharge Q (cfs/in.) | 0 | 35 | 75 | 105 | 40 | 0 |

A 2-hr storm produces 1.7 in. of runoff during the first hour followed by 0.8 in. of runoff during the second hour. The peak discharge (ft³/sec) due to this storm is most nearly:

A.  210
B.  239
C.  263
D.  287

## 120

A triangular channel with side-slopes 1V:3H conveys a flow rate 10 ft³/sec. If the longitudinal slope is 0.5% and Manning's $n = 0.015$, the critical velocity in the channel (ft/sec) is most nearly:

A.  2.9
B.  3.4
C.  3.9
D.  4.4

## 121

A trapezoidal channel has bottom width of 2 ft, a longitudinal slope of 0.5% and sides at 1V:3H slopes, as shown. Manning's n is given as 0.020. If the depth of flowing water is 2 ft, the velocity (fps) is most nearly:

A. 5.6
B. 6.8
C. 3.5
D. 3.8

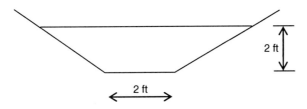

2 ft

2 ft

## 122

A water sample is tested with the following results:

Sample volume filtered = 200 mL

   Mass of crucible and filter paper = 25.439 g
   Mass of dry crucible, filter paper, and solids = 25.645 g
   Mass of crucible, filter paper, and ignited solids = 25.501 g

Sample volume evaporated = 200 mL

   Mass of dry evaporation dish = 275.61 g
   Mass of dry evaporation dish + solids = 275.801 g
   Mass of evaporation dish + ignited solids = 275.645 g

The volatile dissolved solids concentration (mg/L) of the water sample is most nearly:

A. 310
B. 530
C. 780
D. 1,020

## 123

A 12-in. pipe (cast iron, Hazen Williams C = 120) serves as the outfall from the bottom of a large reservoir whose water surface elevation is 145 ft. The average longitudinal slope of the pipe is 0.01 ft/ft. The bottom of the reservoir is at elevation 125 ft and the outfall of the pipe is at elevation 95 ft. Kinematic viscosity of water is $1.217 \times 10^{-5}$ ft²/sec. The flow rate (gal/min) in the pipe is most nearly:

A.  1,000
B.  1,500
C.  2,000
D.  2,500

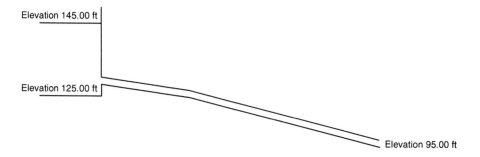

Elevation 145.00 ft

Elevation 125.00 ft

Elevation 95.00 ft

## 124

A watershed (area = 370 acres) is subdivided into 5 distinct land use classifications, as shown in the table below. Precipitation data have been abstracted into the set of intensity-duration-frequency curves shown below. Using the Rational Method, the runoff discharge (ft³/sec) from a 20-year storm with gross rainfall = 5.6 in. is most nearly:

A.  300
B.  600
C.  900
D.  1,200

| Region | Area (acres) | Land Use | Soil Type | Time for Overland Flow (min) | Curve Number | Rational Runoff Coefficient |
|--------|--------------|----------|-----------|------------------------------|--------------|------------------------------|
| A | 80 | Lawns: fair condition | B | 30 | 69 | 0.4 |
| B | 80 | Forest | C | 45 | 45 | 0.2 |
| C | 50 | Paved | B | 15 | 98 | 0.9 |
| D | 90 | Residential: 4 lots per acre | D | 25 | 87 | 0.6 |
| E | 70 | Forest | A | 45 | 35 | 0.2 |

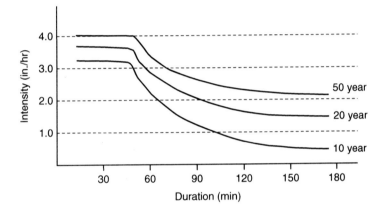

## 125

A parabolic vertical curve is to connect a tangent of +5% to a gradient of −4%. If the PVI is at station 123 + 32.50 and the tangent offset at the PVT is 17.65 ft, the station of the PVC is most nearly:

A. 117 + 12.25
B. 119 + 40.28
C. 119 + 24.21
D. 121 + 36.39

## 126

Spot speed measurements are taken for vehicles passing a given point on an urban road. Results are grouped (5-mph intervals) as shown in the table below.

| Speed Interval (mph) | Frequency |
|---|---|
| >20–25 | 2 |
| >25–30 | 9 |
| >30–35 | 16 |
| >35–40 | 28 |
| >40–45 | 19 |
| >45–50 | 11 |
| >50–55 | 3 |

The 85th percentile speed (mph) is most nearly
A. 40
B. 45
C. 46
D. 50

## 127

If the driver perception time of 2.5 sec is assumed, the braking distance (ft) for a car on a 2% upgrade traveling at 40 mph is most nearly:
A. 145
B. 167
C. 177
D. 307

## 128

A car accelerates uniformly from rest to its peak speed of 70 mph. The acceleration rate is 8 mph/sec. After cruising at maximum speed, the vehicle must brake to a complete stop, decelerating uniformly at 10 mph/sec. If the total distance traveled is exactly 0.5 miles, the average running speed (mph) is most nearly:

A. 48
B. 51
C. 54
D. 57

## 129

A horizontal curve is to be constructed for a paved rural road in a location where the recommended superelevation is 8%. For snow and ice conditions, the recommended value for the side friction coefficient is 0.12. The pavement is 20 ft wide with 4-ft shoulders. The tangents of the proposed curve are perpendicular. The design speed for the roadway is 60 mph. The minimum length of the curve (ft) is most nearly:
A. 1,900
B. 2,200
C. 1,200
D. 1,500

## 130

The coordinates (ft) of the PC and PI for a circular horizontal curve are given below:

  PC:   1232.56 N, 123.32 E
  PI:    509.72 N, 172.11 W

The degree of curve = 4°

The curve deflects to the left.

The coordinates of the PT (ft) are:
A. 256.21 N, 121.72 W
B. 1157.28 S, 811.83 W
C. 1149.62 N, 619.67 E
D. 130.18 S, 275.45 E

## 131

A vertical curve is being designed to connect two grades with $G_1 = +2.1\%$ and $G_2 = -1.5\%$. If the design speed is 65 mph, the minimum length of vertical curve (ft) to satisfy AASHTO safe stopping distance criteria is most nearly:

A. 600
B. 700
C. 800
D. 900

## 132

A stadium hosts an event with an audience of 40,000. Approximately 30% of the audience members are expected to use an adjacent light rail station following the event. It is anticipated that about 90% of the stadium empties in the first hour after the conclusion of the event. A dedicated pedestrian walkway connects the stadium to the light rail station. The effective width of the walkway is 32 ft. Assume that the PHF (based on peak 15-min flow) for the walkway is 0.88. The peak flow rate (ped/min/ft) on the walkway during the first hour is most nearly:

A. 4.8
B. 5.6
C. 6.4
D. 7.2

## 133

An excavator having a capacity of 3 yd³ operates 8 hr in a workday. A single cycle of operation consists of excavation, travel (two-way), and transfer of excavated material. Cycle time for the excavator is 8 min. Efficiency factors are:

Site 0.90
Equipment & Operator 0.72

The daily excavation production (yd³/day) is most nearly:

A. 115
B. 130
C. 145
D. 160

## 134

An activity on arrow network for a project is shown below. Numbers adjacent to arrows are activity durations (weeks). Assume project start date is week 0. The early start date (week) for activity G is:

A.  9
B.  10
C.  11
D.  12

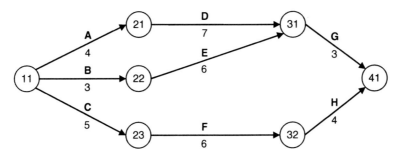

## 135

A concrete mix has proportions 1:1.6:2.6 (cement: sand: coarse aggregate) by weight. The following specifications are given:

Cement                 specific gravity = 3.15
                             94 lb/sack
Sand                    specific gravity = 2.62 (SSD)
Coarse aggregate    specific gravity = 2.65 (SSD)
Added water         5.8 gal/sack cement
Air                   3% (by volume)

The quantity of cement (lb/yd$^3$) is most nearly:

A.  670
B.  650
C.  630
D.  690

## 136

A contractor needs to bring in 4,200 yd³ of select soil to replace unsuitable subgrade material for a new highway. The borrow site is located 2.6 miles away with an average round trip travel/loading/dumping time of 30 min. The soil has a unit weight of 125 lb/ft³ and the dump truck drivers are on 10-hr workdays. The minimum number of 10-ton (net capacity) trucks needed to complete the job within 8 working days is:

A. 4
B. 5
C. 6
D. 7

## 137

A grade beam 3 ft wide × 2 ft deep must be poured around the perimeter of a building whose plan is shown below. The trench excavated for the grade beam has to have 1V:1H side slopes. Excavation is accomplished by a trackhoe, having productivity of 9 yd³/hr. Assume an 8-hr workday.

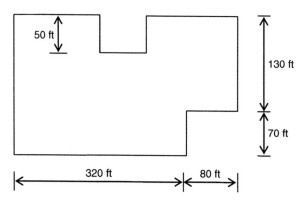

The number of days to complete the excavation activity is most nearly:

A. 4
B. 5
C. 6
D. 7

## 138

A 12-ft-high × 60-ft-long × 12-in.-thick free standing wall is to be constructed using wall forms that are 12-ft high × 20-ft long.

Labor cost for erecting forms = $4.30/ft² (new) and $1.30/ft² (reuse)
Labor cost for dismantling forms = $1.05/ft²
Cost of concrete (assume 10% waste) = $120/yd³
Cost of reinforcement = $25/yd³

The cost of building the wall ($) is most nearly:

A.  7,400
B.  8,100
C.  8,600
D.  9,000

## 139

An embankment is to be constructed by placing fill in a 300-ft-long section of a site. Areas of fill sections 50 ft apart are shown in the following table. The total volume of earthwork (yd³) between stations 12 + 0.00 and 15 + 0.00 is most nearly:

A.  8,060
B.  8,940
C.  16,100
D.  17,880

| Station | Fill Area (ft²) |
|---|---|
| 12 + 00.00 | 456.33 |
| 12 + 50.00 | 563.97 |
| 13 + 00.00 | 702.24 |
| 13 + 50.00 | 1234.98 |
| 14 + 00.00 | 783.92 |
| 14 + 50.00 | 591.94 |
| 15 + 00.00 | 493.34 |

## 140

The tasks within a project, and their duration and cost data are shown below. Both normal completion time and crashed completion time (by allocating additional resources) are shown in the table.

| Activity | Normal Time (weeks) | Normal Cost | Crash Time (weeks) | Crash Cost |
|----------|---------------------|-------------|--------------------|-----------| 
| A | 3 | $3,000 | 2 | $5,000 |
| B | 4 | $4,000 | 2 | $6,000 |
| C | 5 | $5,000 | 3 | $8,000 |
| D | 8 | $5,000 | 6 | $6,000 |
| E | 3 | $3,000 | 2 | $4,000 |
| F | 5 | $4,000 | 3 | $8,000 |

The activity on arrow representation of the project is also shown below.

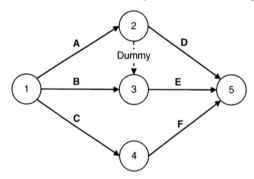

For every week of early completion, the bonus is $1,000, and for every week that the project is late, the penalty is $800. If the revised target completion time is 9 weeks, the net cost for completing the project is most nearly:

A. $22,000
B. $24,000
C. $25,000
D. $26,000

**END OF BREADTH EXAM NO. 2**

**USE THE ANSWER SHEET ON THE NEXT PAGE**

## Breadth Exam No. 2: Answer Sheet

| | | | | |
|---|---|---|---|---|
| 101 | Ⓐ | Ⓑ | **Ⓒ** | Ⓓ |
| 102 | Ⓐ | Ⓑ | Ⓒ | **Ⓓ** |
| 103 | Ⓐ | Ⓑ | Ⓒ | Ⓓ |
| 104 | Ⓐ | Ⓑ | Ⓒ | Ⓓ |
| 105 | Ⓐ | Ⓑ | **Ⓒ** | Ⓓ |
| 106 | Ⓐ | Ⓑ | Ⓒ | Ⓓ |
| 107 | **Ⓐ** | Ⓑ | Ⓒ | Ⓓ |
| 108 | Ⓐ | Ⓑ | Ⓒ | Ⓓ |
| 109 | Ⓐ | **Ⓑ** | Ⓒ | Ⓓ |
| 110 | Ⓐ | **Ⓑ** | Ⓒ | Ⓓ |
| 111 | Ⓐ | Ⓑ | **Ⓒ** | Ⓓ |
| 112 | Ⓐ | Ⓑ | Ⓒ | Ⓓ |
| 113 | Ⓐ | Ⓑ | Ⓒ | Ⓓ |
| 114 | Ⓐ | Ⓑ | Ⓒ | Ⓓ |
| 115 | Ⓐ | Ⓑ | Ⓒ | Ⓓ |
| 116 | Ⓐ | Ⓑ | Ⓒ | Ⓓ |
| 117 | Ⓐ | Ⓑ | **Ⓒ** | Ⓓ |
| 118 | Ⓐ | Ⓑ | Ⓒ | Ⓓ |
| 119 | Ⓐ | Ⓑ | Ⓒ | Ⓓ |
| 120 | Ⓐ | Ⓑ | **Ⓒ** | Ⓓ |

(Bl 1)

| | | | | |
|---|---|---|---|---|
| 121 | **Ⓐ** | Ⓑ | Ⓒ | Ⓓ |
| 122 | Ⓐ | Ⓑ | Ⓒ | Ⓓ |
| 123 | Ⓐ | Ⓑ | Ⓒ | Ⓓ |
| 124 | Ⓐ | Ⓑ | Ⓒ | Ⓓ |
| 125 | Ⓐ | Ⓑ | Ⓒ | Ⓓ |
| 126 | Ⓐ | Ⓑ | Ⓒ | Ⓓ |
| 127 | Ⓐ | Ⓑ | Ⓒ | Ⓓ |
| 128 | Ⓐ | Ⓑ | Ⓒ | Ⓓ |
| 129 | Ⓐ | Ⓑ | Ⓒ | Ⓓ |
| 130 | Ⓐ | Ⓑ | Ⓒ | Ⓓ |
| 131 | Ⓐ | Ⓑ | Ⓒ | Ⓓ |
| 132 | Ⓐ | Ⓑ | Ⓒ | Ⓓ |
| 133 | Ⓐ | Ⓑ | Ⓒ | Ⓓ |
| 134 | Ⓐ | Ⓑ | Ⓒ | Ⓓ |
| 135 | Ⓐ | Ⓑ | Ⓒ | Ⓓ |
| 136 | Ⓐ | Ⓑ | Ⓒ | Ⓓ |
| 137 | Ⓐ | Ⓑ | Ⓒ | Ⓓ |
| 138 | Ⓐ | Ⓑ | Ⓒ | Ⓓ |
| 139 | Ⓐ | Ⓑ | Ⓒ | Ⓓ |
| 140 | Ⓐ | Ⓑ | Ⓒ | Ⓓ |

# 3

# Structural Depth Exam

The following set of questions numbered 201 to 240 is representative of a 4-hour structural depth exam according to the syllabus and guidelines for the Principles and Practice (P&P) of Civil Engineering Examination administered by the National Council of Examiners for Engineering and Surveying (NCEES). Detailed solutions are on page 221.

## 201

A reinforced-concrete beam (15-in. wide × 28-in. deep) has a clear span of 24 ft, as shown in the diagram below. The beam is supported by 10-in.-wide bearing walls, as shown. The superimposed dead load = 2.4 k/ft and the live load = 4 k/ft. Use $f_c' =$ 5,000 psi; $f_y$ = 60,000 psi.

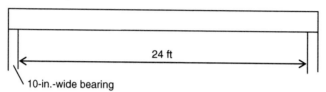

24 ft

10-in.-wide bearing

The required reinforcement (in.²) of the beam is most nearly:

A.   4.45
B.   5.20
C.   6.50
D.   7.65

## 202

For the concrete beam shown in problem 201, the required spacing (in.) of U-shaped no. 4 stirrups at the critical section is most nearly:

A.   3
B.   5
C.   7
D.   9

## 203

A single-story steel frame is shown below. The connection between columns and beam may be assumed to be simple shear connections. Loads shown are factored loads (LRFD). The floor supported by the beam BC does not provide any intermediate lateral support to the beam. Steel used has $F_y$ = 50 ksi and $F_u$ = 70 ksi. Assume $C_b$ = 1.0.

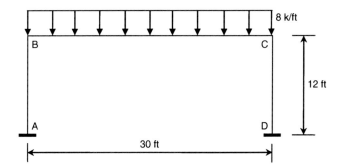

The lightest W-section for beam BC is:
A. W33 × 141
B. W24 × 131
C. W18 × 143
D. W24 × 146

## 204

A steel column ($F_y$ = 36 ksi) has a $KL_x$ = 40 ft and $KL_y$ = 20 ft. Find the smallest satisfactory W12 section that can support an axial load of 200 k dead load + 200 k live load. It is permissible to use either ASD or LRFD.
A. W12 × 79
B. W12 × 87
C. W12 × 96
D. W12 × 106

## 205

A 7.5-in. × 16-in. masonry lintel spans a door opening 5 ft wide. The wall above the lintel is 8 in. thick with a unit weight of 130 lb/ft³. Assume 8-in. bearings on either side of the opening. Assume unit weight of the reinforced lintel to be 140 lb/ft³. The top of the wall carries joists every 16 in. Each joist exerts a vertical reaction of 600 lb on the wall. The maximum bending moment (lb-ft) in the lintel is most nearly:
A. 810
B. 900
C. 990
D. 1,080

(*Continued on next page*)

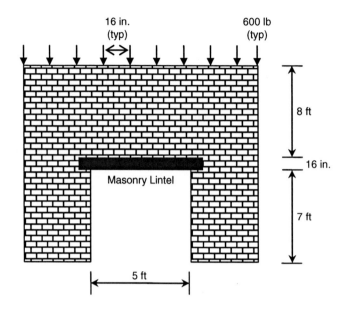

Masonry Lintel

## 206

For the steel beam shown below, the following properties are given: $A_g = 24.7$ in.$^2$; $I_x = 870$ in.$^4$; $I_y = 215$ in.$^4$. The deflection (in.) at point A is most nearly:

A.  0.005
B.  0.010
C.  0.017
D.  0.025

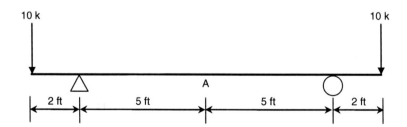

## 207

A composite concrete-deck steel-girder system for a simple span ($L = 70$ ft) bridge is shown below. The beam is a doubly symmetric plate girder ($A_s = 136$ in.$^2$; $I_{xx} = 81,940$ in.$^4$; depth = 60 in.). Slab thickness = 8 in.

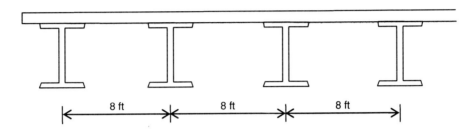

Twenty-eight-day compressive strength of the concrete is 4 ksi. Yield stress for steel is 60 ksi. Load on the deck (including weight of slab plus asphalt overlay and equivalent traffic load) may be taken as 1.8 k/ft$^2$. The maximum bending stress (k/in.$^2$) in the steel girders, due to this unfactored load, is most nearly:

A. 32

B. 36

C. 39

D. 42

## 208

A C10 × 30 is used as a tension member as shown below. Steel used is A572 grade 50. Connection uses ¾-in.-diameter high-strength A325 bolts.

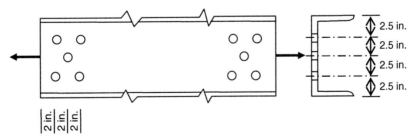

(*Continued on next page*)

|  ASD  |  LRFD  |
|---|---|

**ASD**

The required strength (k) of the beam, as given by AISC-ASD provisions, is most nearly:

A. 205
B. 240
C. 275
D. 305

**LRFD**

The design strength (k) of the member, as given by AISC-LRFD provisions, is most nearly:

A. 310
B. 360
C. 410
D. 460

## 209

The cross section of a short, reinforced-concrete column is shown below. Assume that the eccentricity of the load is 4.0 in. Use $f'_c = 4,000$ psi and $f_y = 60,000$ psi. The design capacity (k) of the column is most nearly:

A. 360
B. 425
C. 465
D. 530

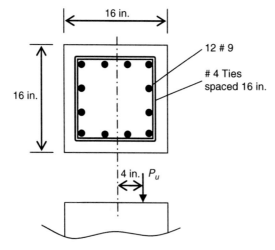

## 210

The propped cantilever beam shown below carries a point load P. The beam is a W18 × 86 section. Assume $F_y$ = 50 ksi and $F_u$ = 70 ksi. The magnitude of the load P (k) that would cause collapse of the beam by plastic hinge formation is most nearly:

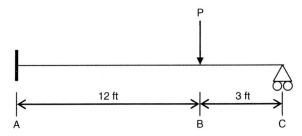

A. 210
B. 256
C. 388
D. 775

## 211

The singly symmetric I-shaped section shown below is fabricated from A36 steel. The plastic moment capacity (k-ft) of the built-up section is most nearly:

A. 1,350
B. 1,650
C. 2,150
D. 2,650

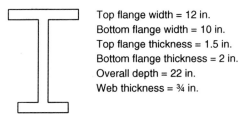

Top flange width = 12 in.
Bottom flange width = 10 in.
Top flange thickness = 1.5 in.
Bottom flange thickness = 2 in.
Overall depth = 22 in.
Web thickness = ¾ in.

## 212

A 3-axle vehicle has a front axle load of 10 k, a drive axle of 28 k, and a rear (trailer) axle of 30 k, as shown below. The maximum bending moment (k-ft) produced by this truck on a 70-ft, simple span is most nearly:

A. 750

B. 840

C. 920

D. 1,050

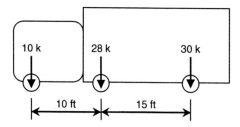

## 213

A floor system consists of a 5-in.-thick reinforced-concrete slab, built monolithically with concrete beams spaced 8 ft, as shown. The beams span 20 ft between simple supports. Use $f_c' = 4,500$ psi and $f_y = 60,000$ psi. The service loads on the slab are:

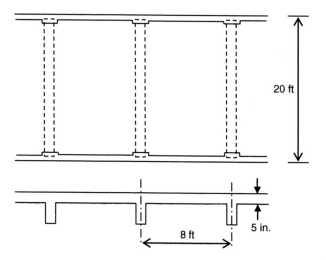

Superimposed dead load = 40 lb/ft²; live load = 85 lb/ft². The flexural reinforcement required (in.²/ft) at the critical section for positive moment is most nearly:

A.  0.32
B.  0.21
C.  0.14
D.  0.11

## 214

A prestressed concrete I-beam section is shown below. The neutral axis is located 13.73 in. below the top fiber. Prestressing tendons (area = 5.90 in.², ultimate stress $f_{pu}$ = 270 ksi, initial prestress = $0.75f_{pu}$, prestress losses = 32 ksi) are located 12.2 in. above the bottom edge of the beam.

The flexural stress (lb/in.²) on the top fiber due to a combination of the prestress force (after all losses), plus a bending moment (due to total DL + LL) = 1,400 k-ft acting on the beam (the gravity loads produce convex curvature on the bottom surface), is most nearly:

A.  845 (tension)
B.  1,630 (compression)
C.  2,145 (compression)
D.  1,230 (tension)

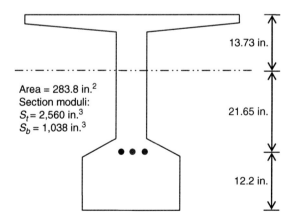

Area = 283.8 in.²
Section moduli:
$S_t$ = 2,560 in.³
$S_b$ = 1,038 in.³

13.73 in.

21.65 in.

12.2 in.

## 215

A steel tension member has a W12 × 72 section and is connected through its flanges using ¾-in.-diameter, high-strength A325 bolts as shown below. $F_y = 36$ ksi; $F_u = 58$ ksi. The nominal strength (k) based on block shear is most nearly:

A.  550
B.  1,090
C.  820
D.  275

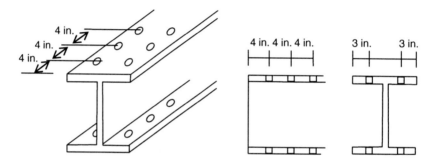

## 216

A 4-story building is supported by a group of vertical piles as shown below. The plan dimensions of the pile cap are 40 ft × 100 ft. Floor height = 15 ft. Based on the pile dimension and the type of soil adjacent to the pile, the following (ultimate) capacities have been established:

Ultimate point bearing capacity = 70 k
Ultimate side friction capacity = 40 k
Weight of each pile = 5.6 k

Loads:

Dead loads:     floors 1-4: 150 k
Roof: 60 k

Live loads:     floors 1-4: 220 k
Roof: 70 k

Lateral loads (wind) shown on elevation.

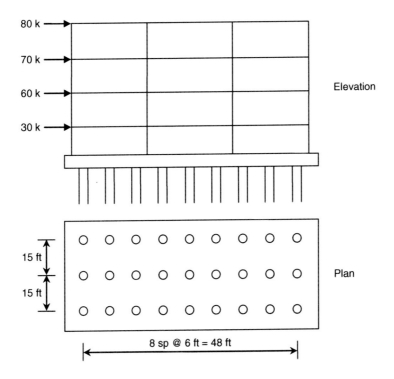

For the case shown (wind parallel to the long plan dimension), the maximum net uplift (k) at a pile is most nearly:

A.  37.8
B.  24.4
C.  13.3
D.  7.8

## 217

A 7-ft × 7-ft reinforced-concrete footing is installed at a depth of 3.5 ft below the ground surface. The unit weight of the soil is 120 lb/ft³. The allowable soil pressure is 3,500 lb/ft². The compressive strength of the concrete is 3,500 lb/in.². The footing experiences the following service loads (concentric) from the 14-in. × 14-in. column:

Dead load    50 k (not including weight of footing)
Live load    90 k

(*Continued on next page*)

The required thickness of the footing (in.), based on punching shear is most nearly:

A. 15

B. 18

C. 21

D. 24

## 218

A bridge deck is carried by two suspension cables. The total deck load is equally transferred to the cables by vertical hangers spaced every 6 ft. The uniformly distributed deck load tributary to each cable is 3.2 k/ft. The maximum tension in the cable (ton) is most nearly:

A. 200

B. 270

C. 360

D. 410

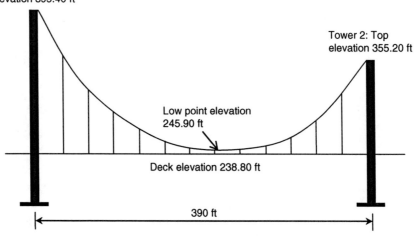

Tower 1: Top
elevation 395.40 ft

Tower 2: Top
elevation 355.20 ft

Low point elevation
245.90 ft

Deck elevation 238.80 ft

390 ft

## 219

A sawn rectangular timber beam (Douglas Fir, 6 in. × 10 in. nominal section) has a simple span (L) of 20 ft and supports a live load of 125 lb/ft and a dead load of 150 lb/ft. The maximum bending stress (psi) is most nearly:

A. 1,400
B. 1,600
C. 1,800
D. 2,000

## 220

A timber post serves as a compression member. The bottom support is adequately restrained against translation and rotation. The top of the column is braced in the weak direction, as shown. The cross section is 2 in. × 6 in. (nominal). The length of the column is 12 ft. Modulus of elasticity $E_{min} = 1.5 \times 10^6$ psi.

The Euler buckling load (k) is most nearly:

A. 1.7
B. 2.5
C. 3.4
D. 5.1

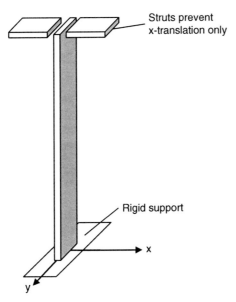

Struts prevent
x-translation only

Rigid support

x

y

## 221

For the plane truss loaded as shown below, the force (k) in member CJ is most nearly:

A. 34 (tension)
B. 34 (compression)
C. 28 (tension)
D. 28 (compression)

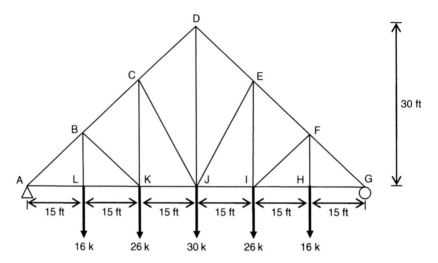

## 222

A circular reinforced-concrete column supports a 300-k dead load and a 350-k live load. The compressive strength of the concrete is 4,000 psi and the yield stress of the reinforcement steel is 60 ksi. The longitudinal reinforcement is laterally confined by a 3/8-in.-diameter spiral with pitch of 3 in. If the maximum reinforcement is used, the required diameter of the column (in.) is most nearly:

A. 13
B. 14
C. 16
D. 17

## 223

What is the maximum shear (k) in a simple span bridge ($L = 125$ ft) due to AASHTO HL-93 loading (unfactored) occupying a single lane?

A. 65
B. 76
C. 107
D. 119

## 224

A system of regularly spaced rectangular timber beams supports a floor as shown below. Assume that modulus of elasticity of the timber is $1.5 \times 10^6$ psi. Beams are 2 in. × 10 in. (1.5 in. × 9.25 in. dressed size). Beams have a simple span of 14 ft. The floor load is 55 psf. If the maximum deflection of the beams is L/240, the allowable spacing S (in.) between beams is most nearly:

A. 16
B. 20
C. 24
D. 28

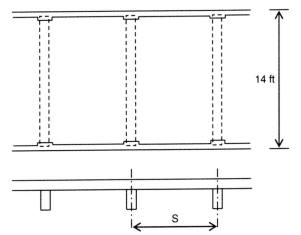

## 225

A W16 × 100 beam has a simple span of 30 ft and carries a uniformly distributed load. Steel grade is A992 grade 50. The compression flange of the beam has lateral support at the supports and at midspan only. Ignore $C_b$.

| ASD | LRFD |
|---|---|
| The maximum load (k/ft) the beam can carry, as given by AISC-ASD provisions, is most nearly: | The maximum factored load (k/ft) the beam can carry, as given by AISC-LRFD provisions, is most nearly: |
| A.  3.3 | A.  5.0 |
| B.  4.0 | B.  6.0 |
| C.  4.7 | C.  7.0 |
| D.  5.2 | D.  8.0 |

## 226

A 16-in. × 20-in. reinforced-concrete ($f'_c = 4,000$ psi; $f_y = 60,000$ psi) column is subjected to the following loads:

$P_D = 300$ k (concentric)
$P_L = 180$ k (eccentricity $e = 4$ in.)

The required area of longitudinal reinforcement (in.²) is most nearly:

A.  3.2
B.  3.8
C.  5.4
D.  6.0

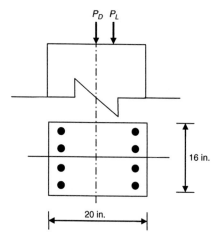

## 227

The plan of a low-rise building is shown. A uniformly distributed wind load, with intensity 200 lb/ft, acts on the building. The lateral load is resisted by the perimeter shear walls. The roof diaphragm is made of wood and may be considered flexible. The unit shear (lb/ft) in wall no. 2 is most nearly:

A.  180
B.  220
C.  260
D.  300

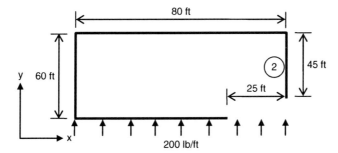

## 228

A36 steel ($F_y = 36$ ksi; $F_u = 58$ ksi) is used for a truss. All loads shown below are service loads. What is the minimum gross area (in.$^2$) needed for the bottom chord? Assume effective net area = $0.75 \times A_g$. (Load is 30% DL + 70% LL).

A.  13.5
B.  17.5
C.  21.5
D.  25.5

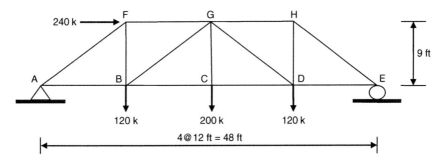

## 229

A hollow steel post (inner diameter = 4.5 in., outer diameter = 5.1 in.) is used to support a rectangular traffic sign as shown below. The sign weighs 750 lb and has no openings. The design wind pressure is 28 psf. Shape factor of post for wind is 0.7.

Ignoring the weight of the sign and the post, the maximum shear stress (k/in.²) in the post is most nearly:

A.  4.4
B.  4.9
C.  2.4
D.  3.6

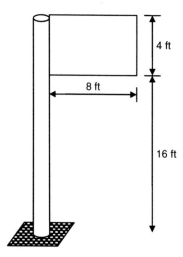

## 230

A steel frame is shown below. The floor load tributary to each frame is 48,000 lb. The fundamental period (sec) of vibration is most nearly:

A.  0.1
B.  0.2
C.  0.3
D.  0.4

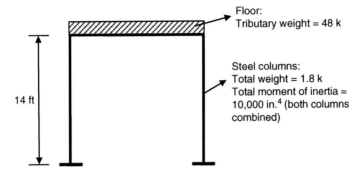

Floor:
Tributary weight = 48 k

Steel columns:
Total weight = 1.8 k
Total moment of inertia =
10,000 in.$^4$ (both columns
combined)

14 ft

## 231

A concrete bridge deck is exposed to freezing and thawing conditions, with moisture
and deicing chemical present continuously. Twenty-eight-day compression strength
$f'_c = 6,000$ psi. Aggregate size = 1 in. The minimum air content (%) is most nearly:

A.  2
B.  3
C.  4
D.  5

## 232

A reinforced-concrete ($f'_c = 4$ ksi; $f_y = 60$ ksi) beam is simply supported over a span
$L = 25$ ft. The beam carries uniformly distributed loads, $w_{DL} = 2$ k/ft, $w_{LL} = 4$ k/ft,
over its length. If the width of the beam is 15 in., the minimum satisfactory depth
(in.) is most nearly:

A.  23
B.  26
C.  29
D.  32

## 233

A doubly symmetric I-section is shown below. Determine if the given section is compact, non-compact, or slender. $F_y = 50$ ksi.

A. Compact web only
B. Non-compact flange only
C. Non-compact (as a whole section)
D. Compact (as a whole section)

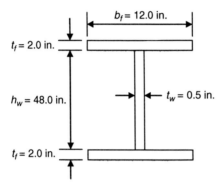

## 234

An elevated water tank is supported by 4 tower legs as shown below. The empty tank weighs 6 k and the full tank weighs 300 k. The resultant wind force of 120 k acts at a height of 65 ft as shown. Each tower leg is supported by an isolated square footing. The maximum design uplift force (k) for designing anchor bolts for each footing is most nearly:

A. 55.0
B. 128.5
C. 130.0
D. 258.5

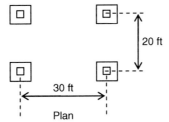

## 235

For the beam shown below, the maximum bending moment (k-ft) is most nearly:

A. 64
B. 72
C. 80
D. 88

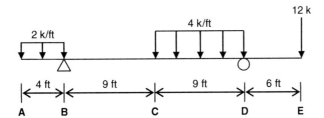

## 236

For the design of a main wind-force resisting system in a building, what is the minimum percentage of the gross wall area that wall openings must occupy in order for a building to be considered an open building?

A. 80
B. 75
C. 90
D. 85

## 237

A two-member truss is loaded with a vertical load of 30 k at node B as shown. The cross-sections of the members AB and BC are: 2 in.$^2$ for AB, 3 in.$^2$ for BC. Modulus of elasticity $E = 29,000$ ksi for both members. The vertical deflection at B (in.) is most nearly:

A. 0.05
B. 0.10
C. 0.15
D. 0.20

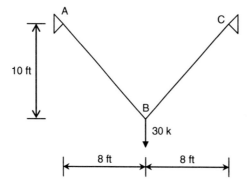

## 238

Computer analysis of a beam ABC shown below results in the following joint moments:

$M_{BA}$ = 65.5 k-ft (clockwise)
$M_{BC}$ = 65.5 k-ft (counterclockwise)
$M_{CB}$ = 209.0 k-ft (clockwise)
$M_{CD}$ = 209.0 k-ft (counterclockwise)

The vertical reaction at B (k) is most nearly:

A.  12.8
B.  19.1
C.  26.4
D.  32.0

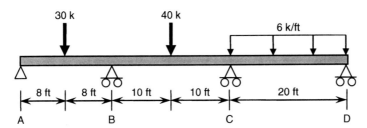

## 239

A reinforced-concrete floor consists of a 5-in.-thick slab cast monolithically with beams as shown below. Twenty-eight-day compressive strength = 4,000 psi. Steel reinforcement is grade 60 deformed bars. Beams have a simple span of 28 ft. The floor live load is 85 psf. Beams are to be designed with tension steel only. The required reinforcement (in.²) of the tensile reinforcement is most nearly:

A.  2.8
B.  3.8
C.  4.8
D.  5.8

*(Continued on next page)*

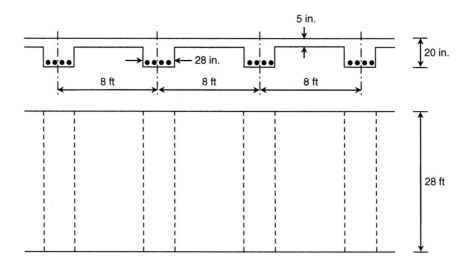

## 240

Which of the following statements must be true in order to determine the design wind pressures based on the simplified procedure in ASCE 7?

    I.   The mean roof height $h$ must be less than or equal to 60 ft ($h \leq 60$ ft).

    II.  The building has response characteristics making it subject to vortex shedding.

    III.  The building is not classified as a flexible building.

    IV.  The building does not have response characteristics making it subject to instability due to galloping or flutter.

A.  I

B.  I and II

C.  I, III, and IV

D.  All of the above

### END OF STRUCTURAL DEPTH EXAM

### USE THE ANSWER SHEET ON THE NEXT PAGE

## Structural Depth Exam: Answer Sheet

| | | | | | | | | | |
|---|---|---|---|---|---|---|---|---|---|
| 201 | Ⓐ | Ⓑ | Ⓒ | Ⓓ | 221 | Ⓐ | Ⓑ | Ⓒ | Ⓓ |
| 202 | Ⓐ | Ⓑ | Ⓒ | Ⓓ | 222 | Ⓐ | Ⓑ | Ⓒ | Ⓓ |
| 203 | Ⓐ | Ⓑ | Ⓒ | Ⓓ | 223 | Ⓐ | Ⓑ | Ⓒ | Ⓓ |
| 204 | Ⓐ | Ⓑ | Ⓒ | Ⓓ | 224 | Ⓐ | Ⓑ | Ⓒ | Ⓓ |
| 205 | Ⓐ | Ⓑ | Ⓒ | Ⓓ | 225 | Ⓐ | Ⓑ | Ⓒ | Ⓓ |
| 206 | Ⓐ | Ⓑ | Ⓒ | Ⓓ | 226 | Ⓐ | Ⓑ | Ⓒ | Ⓓ |
| 207 | Ⓐ | Ⓑ | Ⓒ | Ⓓ | 227 | Ⓐ | Ⓑ | Ⓒ | Ⓓ |
| 208 | Ⓐ | Ⓑ | Ⓒ | Ⓓ | 228 | Ⓐ | Ⓑ | Ⓒ | Ⓓ |
| 209 | Ⓐ | Ⓑ | Ⓒ | Ⓓ | 229 | Ⓐ | Ⓑ | Ⓒ | Ⓓ |
| 210 | Ⓐ | Ⓑ | Ⓒ | Ⓓ | 230 | Ⓐ | Ⓑ | Ⓒ | Ⓓ |
| 211 | Ⓐ | Ⓑ | Ⓒ | Ⓓ | 231 | Ⓐ | Ⓑ | Ⓒ | Ⓓ |
| 212 | Ⓐ | Ⓑ | Ⓒ | Ⓓ | 232 | Ⓐ | Ⓑ | Ⓒ | Ⓓ |
| 213 | Ⓐ | Ⓑ | Ⓒ | Ⓓ | 233 | Ⓐ | Ⓑ | Ⓒ | Ⓓ |
| 214 | Ⓐ | Ⓑ | Ⓒ | Ⓓ | 234 | Ⓐ | Ⓑ | Ⓒ | Ⓓ |
| 215 | Ⓐ | Ⓑ | Ⓒ | Ⓓ | 235 | Ⓐ | Ⓑ | Ⓒ | Ⓓ |
| 216 | Ⓐ | Ⓑ | Ⓒ | Ⓓ | 236 | Ⓐ | Ⓑ | Ⓒ | Ⓓ |
| 217 | Ⓐ | Ⓑ | Ⓒ | Ⓓ | 237 | Ⓐ | Ⓑ | Ⓒ | Ⓓ |
| 218 | Ⓐ | Ⓑ | Ⓒ | Ⓓ | 238 | Ⓐ | Ⓑ | Ⓒ | Ⓓ |
| 219 | Ⓐ | Ⓑ | Ⓒ | Ⓓ | 239 | Ⓐ | Ⓑ | Ⓒ | Ⓓ |
| 220 | Ⓐ | Ⓑ | Ⓒ | Ⓓ | 240 | Ⓐ | Ⓑ | Ⓒ | Ⓓ |

# 4

# Geotechnical Depth Exam

The following set of questions numbered 301 to 340 is representative of a 4-hour geotechnical depth exam according to the syllabus and guidelines for the Principles and Practice (P&P) of Civil Engineering Examination administered by the National Council of Examiners for Engineering and Surveying (NCEES). Detailed solutions are on page 249.

## 301

Results from a standard Proctor compaction test on 6 soil samples from a borrow pit are tabulated in the table below. The natural moisture content of the excavated material is 12%. The fill location requires 1.5 million yd³ of soil compacted to a minimum 90% of the maximum Proctor dry density.

| Sample | Net Weight of Soil (lb) | Moisture Content (%) |
|--------|-------------------------|----------------------|
| 1      | 3.24                    | 12                   |
| 2      | 3.70                    | 14                   |
| 3      | 3.95                    | 16                   |
| 4      | 4.21                    | 18                   |
| 5      | 3.90                    | 20                   |
| 6      | 3.40                    | 22                   |

The total volume of borrow soil that must be excavated (yd³) is most nearly:

A.  1.72 million
B.  1.65 million
C.  1.53 million
D.  1.42 million

## 302

Undrained triaxial tests are conducted on a sample of silty clay. The following results are obtained:

Sample diameter = 2 in.
Sample length = 4 in.
Radial stress = 18 psi
Added axial load at failure = 158.2 lb
Pore pressure at failure = 5.6 psi

If an identical sample is tested in a drained test in which the radial stress is increased to 36 psi, the expected axial load at failure (lb) is most nearly:

A. 290
B. 240
C. 190
D. 160

## 303

A wall footing is embedded 30 in. in a sand layer as shown below. Footing width = 4 ft. Superstructure loads are:

Concentric load = 10 k/ft
Moment = 9 k-ft/ft

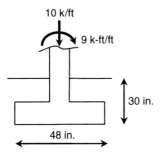

The maximum soil pressure (lb/ft²) under the footing is most nearly:
A. 4,750
B. 5,875
C. 6,000
D. 6,750

## 304

A reinforced-concrete cantilever-retaining wall is shown below. The friction angle between the wall footing and the soil is 20°. The factor of safety for overturning of the wall is most nearly:

(*Continued on next page*)

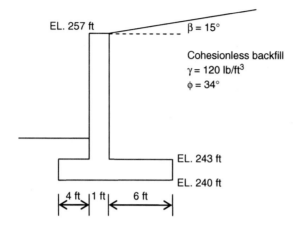

EL. 257 ft

β = 15°

Cohesionless backfill
γ = 120 lb/ft³
φ = 34°

EL. 243 ft

EL. 240 ft

4 ft   1 ft   6 ft

A. 1.2
B. 2.5
C. 3.4
D. 4.8

## 305

For a site, the design earthquake has the following parameters:

Recurrence interval = 540 years
Richter's magnitude = 7.2
Cyclic stress ratio = 0.23

The surface soil layer is a 30-ft-deep bed of fine sand with the following characteristics:

Unit weight $\gamma = 124$ lb/ft³
Angle of internal friction = 34°
Relative density = 0.85

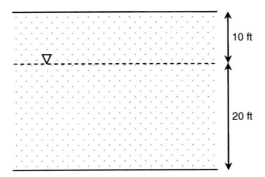

The following results are produced by a laboratory test on a similar soil sample:

Relative density = 95%
Shear stress to cause liquefaction = 1,200 lb/ft²

The factor of safety for liquefaction, at a depth of 30 ft, is most nearly:

A. 1.6
B. 1.9
C. 2.2
D. 2.5

## 306

A 4-story building is supported by a group of vertical piles as shown below. The plan dimensions of the pile cap are 40 ft × 100 ft. Floor height = 15 ft. Based on the pile dimension and the type of soil adjacent to the pile, the following (ultimate) capacities have been established:

Ultimate point bearing capacity = 70 k
Ultimate side friction capacity = 40 k
Weight of each pile = 5.6 k

Loads:

Dead loads:    Floors 1–4: 150 k
               Roof: 60 k

Live loads:    Floors 1–4: 220 k
               Roof: 70 k

Lateral loads (wind) shown on elevation.

(*Continued on next page*)

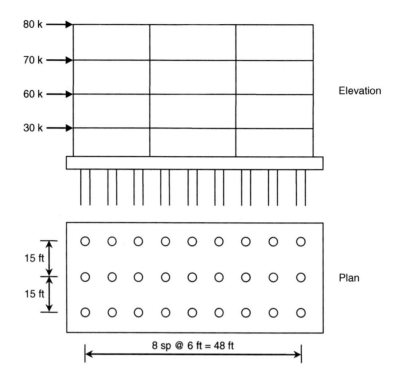

For the case shown (wind parallel to the long plan dimension), the maximum net uplift (k) at a pile is most nearly:

A.  8
B.  13
C.  24
D.  38

## 307

The soil profile shown below consists of a sandy fill layer overlying a layer of normally consolidated clay with the soil properties shown. The soil below the elevation of −20 ft is stiff clay.

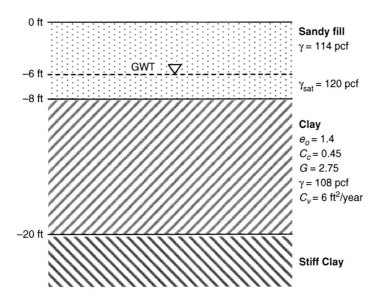

The GWT is lowered to the top of the clay layer and a mat foundation (bottom of mat at depth of 42 in. below ground) of plan dimensions 190 ft × 250 ft is constructed. Total vertical (concentric) load on the mat = 20,000 tons.

The settlement caused by the consolidation of the clay layer (in.) immediately underlying the sandy fill is most nearly

A.  5.0

B.  6.0

C.  7.5

D.  9.0

## 308

For the situation described in problem 307, the time (years) for 80% consolidation to occur is most nearly:

A.  6

B.  9

C.  14

D.  18

## 309

A sample of wet soil weighs 1,331.5 g. The sample is coated with wax (specific gravity = 0.9) and then weighed to be 1,368.2 g. The wax-coated sample is then completely immersed in water and found to weigh 593.4 g. Water content of the soil sample is 15.2% and specific gravity of solids is 2.70. The degree of saturation (%) of the soil sample is most nearly:

A. 57

B. 52

C. 45

D. 37

## 310

Results from a sand cone test are listed below:

| | |
|---|---|
| Net weight of soil obtained from the test hole: | 5.52 lb |
| Moisture content of soil from test hole: | 19% |
| Unit weight of dry test sand: | 88.2 lb/ft³ |
| Initial weight of sand cone apparatus filled with test sand: | 13.75 lb |
| Final weight of sand cone apparatus after sand fills test hole: | 10.24 lb |

A standard Proctor test conducted on the soil resulted in:

| | |
|---|---|
| Maximum dry unit weight | 126.3 lb/ft³ |
| Optimum moisture content | 17.5% |

The in-place percent compaction is most nearly

A. 92

B. 97

C. 114

D. 133

## 311

A dam of length 120 ft is constructed of an impermeable soil. The dam overlays a sand bed of thickness 6.5 ft, as shown below. The sand has the following characteristics:

Unit weight = 125 lb/ft³

Void ratio = 0.45

Water content = 18%

Hydraulic conductivity = $1 \times 10^{-4}$ ft/sec
Scour velocity for sand = 8 in./hr

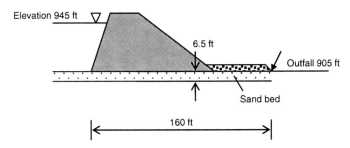

The factor of safety for scouring in the sand drain is most nearly:
A.  0.9
B.  1.8
C.  2.3
D.  2.9

## 312

The liner for a landfill cell has the cross section shown below. It consists of a 12-in.-thick clay layer sandwiched between sand layers of thickness 6 in. Permeability of the sand is approximately 100 times that of the clay. Water collects to a depth of 6-in. on top of the clay liner and flows through the 12-in. layer into a leachate collection system, which is drained to maintain a water level at the bottom of the clay layer. The time (days) for a contaminant to flow through the clay liner is most nearly:
A.  1,000
B.  1,500
C.  2,500
D.  3,500

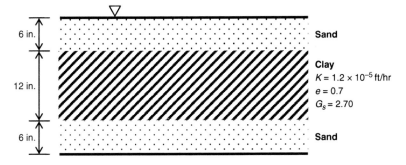

## 313

The following data is given for a soil sample:

| Sieve Analysis: | | Atterberg Tests: |
|---|---|---|

| Sieve size | Percent Retained |
|---|---|
| No. 4 | 8 |
| No. 10 | 10 |
| No. 20 | 12 |
| No. 40 | 21 |
| No. 100 | 15 |
| No. 200 | 8 |

Atterberg Tests:

Liquid Limit = 43
Plastic Limit = 21

The USCS soil classification is:
A. SM
B. GW
C. GP
D. SC

## 314

A 5 ft × 5 ft square footing carries a concentric load of 100 k. The depth of the footing is 42 in. The vertical stress increase (lb/ft$^2$) at a point 10 ft below one of the corners of the footing is most nearly:
A. 150
B. 350
C. 600
D. 1,200

## 315

A pile group is as shown below (all dimensions are between pile centers). Piles are driven, precast concrete piles of diameter 24 in. The total concentric superstructure load (including the weight of the pile cap) is 125 tons. The settlement (in.) due to the consolidation of the clay layer is most nearly:

A. 4.0
B. 2.5
C. 5.0
D. 6.0

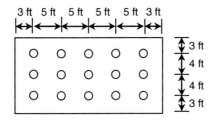

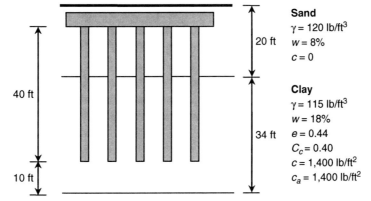

## 316

A temporary slope will be excavated in clay as shown in the figure below. The height of the embankment is 20 ft and the slope angle is 30°. A presumed failure surface (base circle), which is tangent to a firm layer, is shown (dashed line). The firm layer is at a depth of 30 ft below the base of the slope. For this failure surface, the factor of safety for slope stability is most nearly:

A. 1.5
B. 1.8
C. 2.1
D. 2.4

(*Continued on next page*)

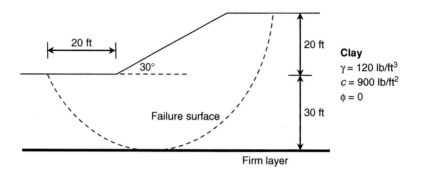

## 317

To construct an embankment, it will take 500,000 cu yd of soil compacted to at least 95% of the standard Proctor maximum dry density (MDD). In a proposed borrow pit, the soil has an in-situ total density $\gamma = 108$ pcf at a water content of 20%. The plan is to transport the soil using trucks having a 10 cu yd capacity of the borrow soil at a total density of 115 pcf and water content of 20%. If the standard Proctor MDD for this soil equals 95 pcf and the optimum moisture content equals 23%, the number of truck loads required to construct the embankment, assuming no loss of soil, is most nearly:

A. 42,000 truck loads
B. 47,000 truck loads
C. 52.000 truck loads
D. 57,000 truck loads

## 318

An anchored bulkhead is used to retain soil behind a vertical trench, as shown below. The dredge side elevation is 120 ft and the water table is at elevation 134 ft. The anchor is located 1 ft above the water table. The bulkhead is driven to a depth of 105 ft. The total lateral pressure (lb/ft) on the bulkhead due to the active soil is most nearly:

A. 15,200
B. 25,400
C. 41,300
D. 45,700

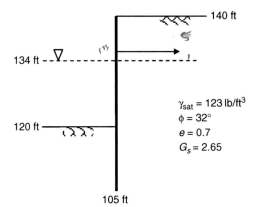

## 319

A 20-ft-wide × 24-ft-deep trench in a clayey soil is braced as shown below. Longitudinal spacing of the struts is 8 ft. The axial load in strut $S_1$ (k) is most nearly:

A. 10
B. 15
C. 20
D. 25

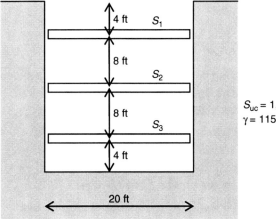

## 320

A 5-ft-diameter concrete pipe will be installed in an 8-ft-wide ditch and covered with 14 ft of compacted gravelly sand backfill with $\gamma = 125$ lb/ft$^3$ and $\phi = 32°$.

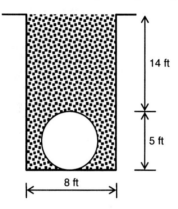

Using standard procedures for the design of buried pipes, the total load on the pipe (lb/ft) is most nearly:

A. 4,500
B. 6,000
C. 8,000
D. 11,000

## 321

The footing for a bridge is supported by a group of 8 precast concrete point-bearing piles, as shown below. Each pile has an allowable load of 30 tons. Service loads from the superstructure transmitted to the pile group are:

  Axial load: DL + LL (including weight of pile cap) = 350 k
  Moment: DL + LL = 300 ft-k
  28-day compression strength of concrete $f'_c = 4,000$ psi
  Yield strength of reinforcement steel $f_y = 60,000$ psi

The minimum required spacing B (ft) of the piles is most nearly:

A. 2 ft 6 in.
B. 3 ft 0 in.
C. 3 ft 6 in.
D. 4 ft 0 in.

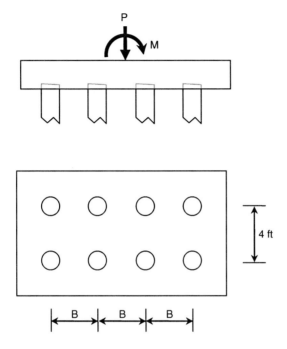

## 322

A reinforced-concrete dam has a sheet pile cutoff wall parallel to its upstream face, as shown below. The elevation of the reservoir is 213.45 ft A.S.L. The underlying soil is silty sand with hydraulic conductivity (K) of 200 ft/day. The dam is 150 ft long. The flow net has been constructed and is shown. The total seepage loss (ft³/sec) under the dam is most nearly:

A. 0.013

B. 0.1

C. 2.0

D. 13.7

(*Continued on next page*)

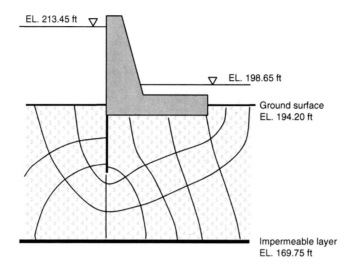

## 323

Three cylindrical samples (diameter = 4 in., thickness = 2 in.) of the same soil are tested in direct shear using a split mold and subjected to a shear force pair, as shown in the figure below. The top and bottom surfaces of the sample are capped with porous stone-loading plates. The normal force N (lb) is varied for each sample. The horizontal shear force F (lb) required to cause shear failure is also recorded. These values are summarized in the table after the figure. The cohesion of the soil (lb/ft²) is most nearly:

A. 370
B. 575
C. 760
D. 980

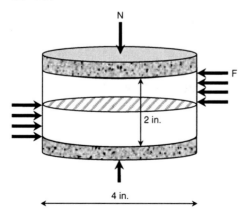

| Sample | N (lb) | F (lb) |
|--------|--------|--------|
| 1 | 120 | 78 |
| 2 | 160 | 93 |
| 3 | 220 | 116 |

## 324

The reinforced-concrete cantilever-retaining wall pictured below has been backfilled with clean granular soil, is free to rotate, and has good drainage so that hydrostatic pressure will not build up behind the wall. The backfill, foundation, and concrete have the following properties:

Concrete: $\gamma_c = 150$ pcf
Granular backfill $\gamma = 118$ pcf
$K_A = 0.3$
$K_P = 3.3$
$\mu$ (base of wall) $= 0.5$

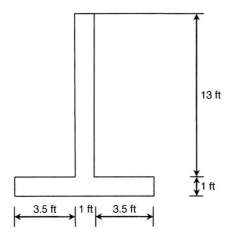

Assuming no wall friction, the factor of safety against sliding is most nearly:

A. 2.5

B. 4.25

C. 0.25

D. 1.25

## 325

A 10 ft × 10 ft reinforced-concrete footing is installed at a depth of 5 ft below the ground surface. The unit weight of the soil is 100 lb/ft$^3$. The allowable soil pressure is 3,500 lb/ft$^2$. The compressive strength of the concrete is 3,500 lb/in.$^2$. The footing experiences the following service loads (concentric) from the column:

Dead load     125 k (not including weight of footing)
Live load     175 k

The plan area contributing to two-way punching shear has been calculated as 92.5 ft$^2$. The ultimate two-way punching shear (k) is most nearly:

A.  280
B.  325
C.  450
D.  500

## 326

The soil profile shown below consists of a sandy fill layer overlying a layer of normally consolidated clay with the soil properties shown.

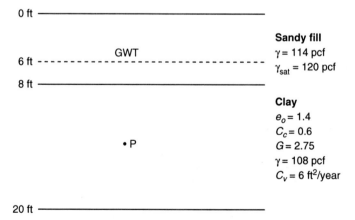

If a building foundation (depth = 4 ft) exerts a stress increase of 670 lb/ft$^2$ at point P located at the center of the clay layer, the settlement caused by the consolidation of the clay layer (in.) is most nearly:

A. 7.5
B. 6.0
C. 4.5
D. 9.0

## 327

A 10-ft square footing embedded 2 ft into a sand profile has a total density, $\gamma_t$, equal to 115 pcf and bearing capacity factors $N_c = 48$, $N_q = 25$, and $N_\gamma = 19$. Neglecting shape factors, if the depth to the water table is 7 ft below the ground surface and the proposed column load equals 750 k, the FS against bearing capacity failure is most nearly:

A. 1.5
B. 1.8
C. 2.2
D. 2.6

## 328

It is proposed to drive a concrete pile using a pile hammer with energy of 50,000 ft-lb. The minimum factor of safety against bearing capacity failure is equal to 6.0. The superstructure load transmitted to the pile is expected to be 40 tons.

The ENR pile driving equation gives the static (ultimate) bearing capacity of a pile based on a pile driving test as:

$$Q_{ult} = \frac{WH}{s+1.0}$$

where  $Q_{ult}$ = ultimate capacity (lb)
  $W$ = weight of the ram (lb)
  $H$ = height the ram falls (in.)
  $S$ = pile set in in. per blow

The required blow count (blows per foot) to achieve a design capacity of 40 tons is most nearly:

A. 12
B. 28
C. 35
D. 50

## 329

Water seeps through a sandy soil layer (20-in.-thick, 30-in.-diameter) and overflows through an outlet maintaining the outfall elevation of 50 in., as shown below. As the water seeps out, the level of water drops in the fine bore tube from an elevation of 85 in. to 75 in., over a 35-min interval. The hydraulic conductivity (ft/hr) of the soil is most nearly:

A.  $1.8 \times 10^{-5}$

B.  $4.3 \times 10^{-4}$

C.  $5.7 \times 10^{-5}$

D.  $6.5 \times 10^{-4}$

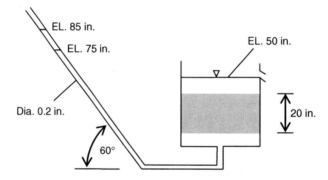

## 330

A reinforced-concrete cantilever-retaining wall is shown below. The total width of the wall footing is 9 ft. Excluding the passive soil, the factor of safety against overturning is most nearly:

A.  3.7

B.  4.7

C.  2.1

D.  3.1

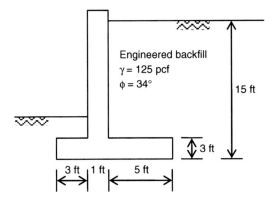

## 331

A clay embankment has a slope of 3H:1V and a height of 18 ft, as shown. The unit weight of the clay is 125 lb/ft³ and the unconfined compression strength is 1,600 lb/ft². The factor of safety for slope stability (Taylor) is most nearly:

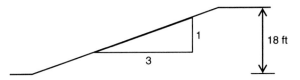

A.   0.8
B.   1.6
C.   1.8
D.   2.0

## 332

The following results are listed for a soil:

| | |
|---|---|
| Percent passing No. 4 sieve (4.75 mm) | 80 |
| Percent passing No. 10 sieve (2.00 mm) | 60 |
| Percent passing No. 40 sieve (0.425 mm) | 30 |
| Percent passing No. 200 sieve (0.075 mm) | 10 |
| Liquid limit | 31 |
| Plastic limit | 25 |

*(Continued on next page)*

What is the classification of the soil according to the Unified Soil Classification System (USCS)?

A. SM
B. SP
C. SW-SC
D. SW-SM

### 333

A 5-ft × 5-ft square footing carries a concentric load of 100 k. The depth of the footing is 3 ft. The vertical stress increase ($lb/ft^2$) at a point 4 ft below and 6 ft laterally offset from the center of the footing is most nearly:

A. 100
B. 150
C. 200
D. 250

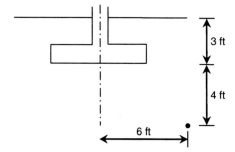

### 334

A trench is excavated in sandy soil to a depth of 20 ft. The trench is supported by timber planking as shown below. The horizontal earth pressure behind the sheet piles is calculated at 500 $lb/ft^2$. The vertical sheet piles are supported by longitudinal wales (vertical spacing = 4 ft) as shown. If the allowable bending stress in the timber is 1,400 $lb/in.^2$, the required thickness (in.) of the planks is most nearly:

A. 1.0
B. 1.5
C. 1.75
D. 2.0

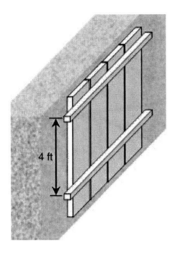

## 335

The cyclic stress ratio for a site subjected to the design earthquake is 0.21. The typical soil stratification at the site is shown below. The unconfined compression strength (dynamic) of the clay is 2,000 lb/ft$^2$.

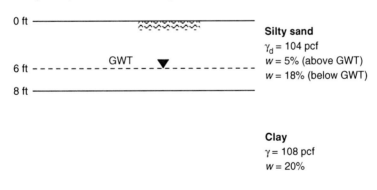

The factor of safety for shear failure at the top of the clay layer is most nearly:

A. 1.5
B. 2.5
C. 4.5
D. 6.0

## 336

The investigation of the stability of a sloped embankment using the method of slices is shown below.

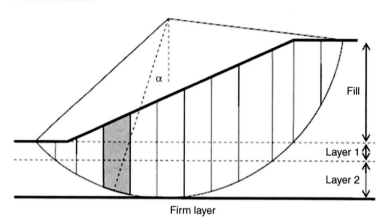

The appropriate expression for the factor of safety for slope stability is:

A. $FS = \dfrac{\Sigma W \sin\alpha + \Sigma cL}{\Sigma(W\cos\alpha - uL)\tan\phi}$

B. $FS = \dfrac{\Sigma(W\cos\alpha - uL)\tan\phi + \Sigma cL}{\Sigma W \sin\alpha}$

C. $FS = \dfrac{\Sigma W \sin\alpha - \Sigma cL}{\Sigma(W\cos\alpha - uL)\tan\phi}$

D. $FS = \dfrac{\Sigma(W\cos\alpha - uL)\tan\phi}{\Sigma W \sin\alpha - \Sigma cL}$

## 337

A 30-ft-deep aquifer is confined between two impermeable layers of rock as shown below. The elevation of the piezometric surface is at 20 ft above the top of the aquifer. A 9-in.-diameter well is used to establish a steady state pumping rate of 2,000 gal/min. The hydraulic conductivity of the soil in the aquifer is 1,000 ft/day. Observation wells 1 and 2 are drilled at radial distance 30 ft and 180 ft from the centerline of the pumping well, as shown. The drawdown of the piezometric surface at observation well no. 1 is 4.5 ft. The drawdown of the piezometric surface (ft) at observation well no. 2 is most nearly:

A.  2.1
B.  2.8
C.  3.2
D.  3.8

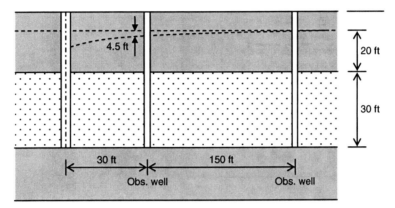

## 338

The square footing shown below supports a concentric column load. Minimum factor of safety based on ultimate bearing capacity is 2.8. The maximum column load (k) is most nearly:

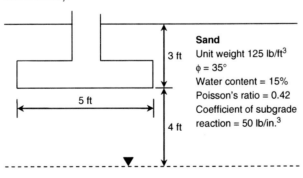

A.  230
B.  245
C.  260
D.  275

## 339

A pile group is as shown below. The total superstructure load transmitted to the pile cap is 450 tons. Each pile (24-in.-diameter) has capacity of 95 tons. The pile group efficiency (%) is most nearly:

A. 65

B. 80

C. 125

D. 175

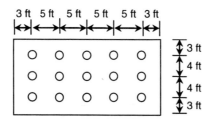

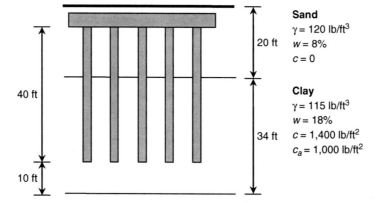

## 340

The soil profile shown below consists of a silty sand layer overlying a 12-ft-thick layer of medium sand. A standard penetration test is conducted with a split spoon sampler, which is driven through an 18-in. penetration from depth of 10 ft to 11 ft 6 in. The number of blows to drive the sampler through 6-in. penetration intervals are 12, 17, and 22, respectively.

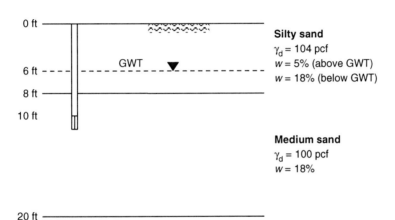

The corrected standard penetration resistance (N-value) is most nearly:

A. 31
B. 39
C. 57
D. 62

**END OF GEOTECHNICAL DEPTH EXAM**

**USE THE ANSWER SHEET ON THE NEXT PAGE**

# Geotechnical Depth Exam: Answer Sheet

| 301 | Ⓐ | Ⓑ | Ⓒ̶ | Ⓓ | ✗ | 321 | Ⓐ | Ⓑ | Ⓒ | Ⓓ |
|-----|---|---|---|---|---|-----|---|---|---|---|
| 302 | Ⓐ | Ⓑ | Ⓒ | Ⓓ | | 322 | Ⓐ | Ⓑ | Ⓒ | Ⓓ |
| 303 | Ⓐ | Ⓑ | Ⓒ | Ⓓ | | 323 | Ⓐ | Ⓑ | Ⓒ | Ⓓ |
| 304 | Ⓐ | Ⓑ | Ⓒ | Ⓓ | | 324 | Ⓐ | Ⓑ | Ⓒ | Ⓓ |
| 305 | Ⓐ | Ⓑ | Ⓒ | Ⓓ | | 325 | Ⓐ | Ⓑ | Ⓒ | Ⓓ |
| 306 | Ⓐ | Ⓑ | Ⓒ | Ⓓ | | 326 | Ⓐ | Ⓑ | Ⓒ | Ⓓ |
| 307 | Ⓐ | Ⓑ | Ⓒ | Ⓓ | | 327 | Ⓐ | Ⓑ | Ⓒ | Ⓓ |
| 308 | Ⓐ | Ⓑ | Ⓒ | Ⓓ | | 328 | Ⓐ | Ⓑ | Ⓒ | Ⓓ |
| 309 | Ⓐ | Ⓑ | Ⓒ | Ⓓ | | 329 | Ⓐ | Ⓑ | Ⓒ | Ⓓ |
| 310 | Ⓐ | Ⓑ | Ⓒ | Ⓓ | | 330 | Ⓐ | Ⓑ | Ⓒ | Ⓓ |
| 311 | Ⓐ | Ⓑ | Ⓒ | Ⓓ | | 331 | Ⓐ | Ⓑ | Ⓒ | Ⓓ |
| 312 | Ⓐ | Ⓑ | Ⓒ | Ⓓ | | 332 | Ⓐ | Ⓑ | Ⓒ | Ⓓ |
| 313 | Ⓐ | Ⓑ | Ⓒ | Ⓓ | | 333 | Ⓐ | Ⓑ | Ⓒ | Ⓓ |
| 314 | Ⓐ | Ⓑ | Ⓒ | Ⓓ | | 334 | Ⓐ | Ⓑ | Ⓒ | Ⓓ |
| 315 | Ⓐ | Ⓑ | Ⓒ | Ⓓ | | 335 | Ⓐ | Ⓑ | Ⓒ | Ⓓ |
| 316 | Ⓐ | Ⓑ | Ⓒ | Ⓓ | | 336 | Ⓐ | Ⓑ | Ⓒ | Ⓓ |
| 317 | Ⓐ | Ⓑ | Ⓒ | Ⓓ | | 337 | Ⓐ | Ⓑ | Ⓒ | Ⓓ |
| 318 | Ⓐ | Ⓑ | Ⓒ | Ⓓ | | 338 | Ⓐ | Ⓑ | Ⓒ | Ⓓ |
| 319 | Ⓐ | Ⓑ | Ⓒ | Ⓓ | | 339 | Ⓐ | Ⓑ | Ⓒ | Ⓓ |
| 320 | Ⓐ | Ⓑ | Ⓒ | Ⓓ | | 340 | Ⓐ | Ⓑ | Ⓒ | Ⓓ |

# 5

# Water Resources & Environmental Depth Exam

The following set of questions numbered 401 to 440 is representative of a 4-hour water resources & environmental depth exam according to the syllabus and guidelines for the Principles and Practice (P&P) of Civil Engineering Examination administered by the National Council of Examiners for Engineering and Surveying (NCEES). Detailed solutions are on page 271.

## 401

A water sample yields the following results:

| | |
|---|---|
| $Ca^{++}$ | 60.0 mg/L |
| $Mg^{++}$ | 21.2 mg/L |
| $Fe^{++}$ | 2.2 mg/L |
| Na+ | 5.5 mg/L |
| $HCO_3^-$ | 221.3 mg/L |
| $Cl^-$ | 33.5 mg/L |
| $NO_3^-$ | 21 mg/L |
| Turbidity | 1.3 NTU |
| Odor | 2.7 TON |
| Total coliform | 1.7 MPN |
| TDS | 437 mg/L |

Which of the following contaminants exceed the EPA's primary drinking water standards?

A. Nitrate and turbidity

B. Turbidity, total dissolved solids, and total coliform

C. Odor, total coliform, and TDS

D. Total coliform and turbidity

## 402

A daily pattern (typical) of wastewater flow rates influent into a wastewater treatment plant is summarized in the following table. The flow undergoes hydraulic stabilization by first passing through an equalization tank. The minimum volume (gal) of the equalization tank is most nearly:

A. $2.6 \times 10^5$

B. $6.3 \times 10^5$

C. $1.5 \times 10^6$

D. $2.9 \times 10^6$

| Time Period (hr) | Average Inflow Rate (ft³/sec) |
|---|---|
| 00:00–02:00 | 18.3 |
| 02:00–04:00 | 12.4 |
| 04:00–06:00 | 8.7 |
| 06:00–08:00 | 7.3 |
| 08:00–10:00 | 6.4 |
| 10:00–12:00 | 8.9 |
| 12:00–14:00 | 14.5 |
| 14:00–16:00 | 18.9 |
| 16:00–18:00 | 6.5 |
| 18:00–20:00 | 5.6 |
| 20:00–22:00 | 8.3 |
| 22:00–24:00 | 13.2 |

## 403

A 200-mL water sample is filtered through a standard Whatman filter. The filtrate is evaporated at 105°C in an evaporation dish (mass = 45.675 g). Following evaporation, the mass of the evaporation dish + solids = 47.225 g. The dish + solids are then ignited at 550°C, following which the mass of the evaporation dish + solids = 46.201 g.

The following data are also given for the solids retained on the filter paper.

Mass of crucible and filter paper = 25.334 g
Mass of crucible, filter paper, and dry (evaporated at 105°C) solids = 25.645 g
Mass of ignited crucible, filter paper, and solids = 25.501 g

The volatile dissolved solids concentration (mg/L) of the waste sample is most nearly:

A. 1,300
B. 1,800
C. 2,600
D. 5,100

## 404

An industrial plant produces wastewater with the following characteristics:

Flow rate = 2 MGD
Temperature = 38°C
Ultimate BOD = 280 mg/L
Dissolved oxygen = 2.1 mg/L
Lead concentration = 0.5 mg/L

The stream into which the plant plans to discharge its wastewater has the following characteristics:

Flow rate = 30 ft³/sec
Temperature = 14°C
Ultimate BOD = 10 mg/L
Lead concentration = 4 µg/L

If the EPA limit for lead in surface waters is 15 µg/L, the level of lead removal (%) necessary at the plant (before discharging into the stream) is most nearly:

A.  80
B.  75
C.  85
D.  0

## 405

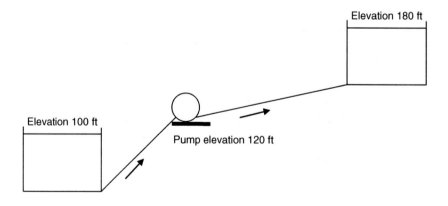

In the above diagram, the pump located at an elevation of 120 ft is used to pump a flow $Q = 3,000$ gpm from a reservoir with surface elevation of 100 ft to another reservoir with surface elevation of 180 ft. The approximate efficiency of the pump is 88%. The characteristics of the pipe system are given below:

| Suction line: | 800-ft length, 18-in. diameter, friction factor = 0.024 |
| | Total minor loss coefficient (valves, bends, etc.) = 5 |
| Discharge line: | 2,500-ft length, 12-in. diameter, friction factor = 0.026 |
| | Total minor loss coefficient (valves, bends, etc.) = 25 |

The required brake horsepower of the pump is most nearly:

A. 100
B. 130
C. 160
D. 200

## 406

A rectangular sharp-crested contracted weir is used to measure flow across a 20-ft-wide open channel. The weir opening is 6 ft wide. The weir head (upstream) = 5.4 ft. The channel discharge (ft³/sec) is most nearly:

A. 835
B. 790
C. 250
D. 205

## 407

The table below shows recorded discharge at a stream monitoring station following a 2-hr storm. The tributary watershed area contributing runoff to the stream has been established as 115 acres.

| Time (hr) | 0 | 1 | 2 | 3 | 4 | 5 | 6 |
|---|---|---|---|---|---|---|---|
| Discharge Q (ft³/sec) | 23 | 84 | 127 | 112 | 75 | 32 | 25 |

The peak stream discharge (ft³/sec) that would be recorded following a 2-hr storm, which produces 1.7 in. of runoff, is most nearly:

A. 90
B. 65
C. 105
D. 127

## 408

A 230-acre watershed can be subdivided into five major parts based on land use and land cover, as shown below:

| Region | Area (acres) | Soil Type | Land Use | Overland Flow Time (min) |
|---|---|---|---|---|
| 1 | 80 | C | Single family homes on ½-acre lots | 25 |
| 2 | 50 | D | Lawns in good condition | 42 |
| 3 | 10 | B | Paved streets and sidewalks | 15 |
| 4 | 50 | C | Grassy areas: fair condition | 34 |
| 5 | 40 | A | Woods: fair condition | 40 |

The NRCS curve number for this drainage basin is most nearly:

A. 55

B. 64

C. 73

D. 82

## 409

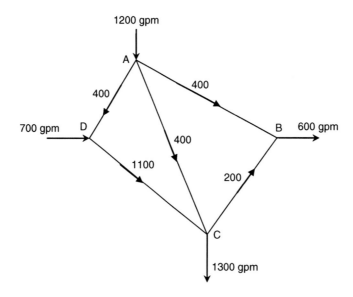

A pipe network is shown. Inflows occur at nodes A and D. Outflows occur at nodes B and C. The presumed (trial) flows are shown on the various links (network is not balanced). The pipe characteristics are shown in the table below:

| Pipe | Length (ft) | Friction Factor | Diameter (in.) |
|------|-------------|-----------------|----------------|
| AB | 600 | 0.020 | 8 |
| AC | 700 | 0.030 | 12 |
| AD | 500 | 0.020 | 6 |
| CD | 400 | 0.025 | 8 |
| BC | 300 | 0.030 | 12 |

After the first iteration of the Hardy Cross method, the adjusted flow (gpm) in pipe AD is most nearly:
A. 80 gpm (A to D)
B. 720 gpm (A to D)
C. 720 gpm (D to A)
D. 80 gpm (D to A)

## 410

The following figure shows manufacturer's curves for a centrifugal pump. Two of these pumps are used in series to pump water from a reservoir with surface elevation of 150 ft to another with surface elevation of 220 ft. All pipes have Hazen-Williams C = 100. Minor losses may be assumed to be 10% of friction losses in the system. The discharge (gpm) at the operating point is most nearly:
A. 800
B. 1,000
C. 1,200
D. 1,400

(*Continued on next page*)

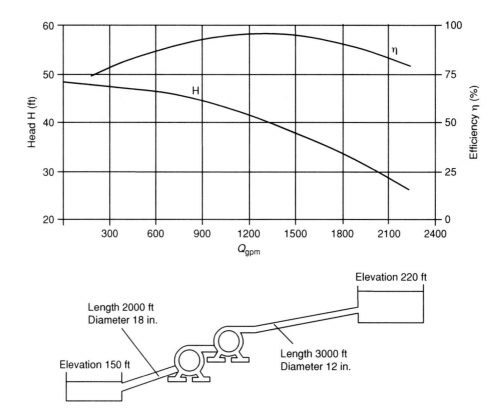

## 411

The flow rate treated at a wastewater treatment plant is 3.2 MGD. The total suspended solids (TSS) concentration in the influent is 800 mg/L. The flow passes through a bank of filters, arranged in parallel.

Filter size is limited to 250 ft$^2$.

The maximum solids load on each filter is 15 lb-TSS/ft$^2$-day.

The maximum surface loading velocity on the filters = 10 ft/hr.

The number of filters needed is most nearly:
A. 6
B. 7
C. 8
D. 9

## 412

The oxygen demand for a wastewater sample at 5 days is 4.5 mg/L, and at 20 days is 8.3 mg/L. The deoxygenation rate constant (base 10, 20°C) is 0.1 day$^{-1}$. The nitrogenous oxygen demand (mg/L) of the sample is most nearly:

A.  3.8
B.  6.6
C.  0.0
D.  1.8

## 413

A stormwater detention pond is approximated as 1 acre in plan area with nearly vertical sides. The pond has a 5-ft-wide sharp-crested weir whose crest elevation is 125.0 ft above sea level. At a particular instant, the inflow to the pond was measured as 50 cfs and the surface elevation of the pond was 126.4 ft. Assuming the inflow rate stays constant, the outflow (cfs) from the pond 5 min later is most nearly:

A.  30
B.  49
C.  57
D.  65

## 414

An orifice meter (orifice diameter = 2 in.) is inserted in a 4-in.-diameter cast iron pipe carrying water at 15°C. The pressure difference across the meter is 30 lb/in.$^2$. The orifice has the following coefficients: $C_c = 0.9$, $C_v = 0.95$. The discharge (gal/min) through the pipe is most nearly:

A.  420
B.  570
C.  760
D.  920

## 415

A dam of length 120 ft is constructed of an impermeable soil. The dam overlays a sand bed of thickness 6.5 ft as shown below. The sand has the following characteristics:

Unit weight = 125 lb/ft³
Void ratio = 0.45
Water content = 18%
Hydraulic conductivity = $1 \times 10^{-4}$ ft/sec
Scour velocity for sand = 8 in./hr

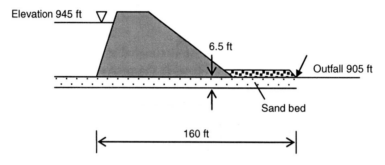

The factor of safety for scouring in the sand drain is most nearly:

A. 0.9
B. 1.8
C. 2.3
D. 2.9

## 416

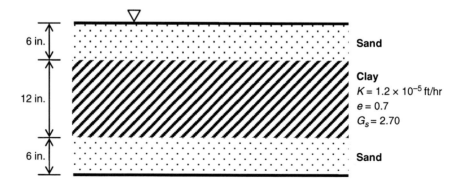

The liner for a landfill cell has the cross section shown. It consists of a 12-in.-thick clay layer sandwiched between sand layers of thickness 6 in. Permeability of the sand is approximately 100 times that of the clay. Water collects to a depth of 6 in. on top of the clay liner and flows through the 12-in. layer into a leachate collection system, which is drained to maintain a water level at the bottom of the clay layer. The time (days) for a contaminant to flow through the clay liner is most nearly:

A. 1,000

B. 1,500

C. 2,500

D. 3,500

## 417

A 36-in.-diameter reinforced-concrete pipe under a roadway serves as a culvert operating under inlet control. The following data apply to the culvert:

Upstream invert: Elevation 325.0

Downstream invert: Elevation 322.5

Length of culvert: 145 ft

Entrance condition: Headwall with square edge

Roadway elevation: 337.8 ft

Culvert flow rate due to 100-year flood event: 100 cfs

The minimum vertical clearance (ft) between the roadway and the 100-year flood elevation is most nearly:

A. 5.2

B. 2.2

C. 4.8

D. 7.8

## 418

The following pipe system shown circulates water at 65°F. Flow enters the network at B and leaves it at node G. The table shows hydraulic characteristics of all pipes in the network. When the flow rate in and out of the network is 300 gal/min, the total head loss between nodes A and H is 70 ft.

(*Continued on next page*)

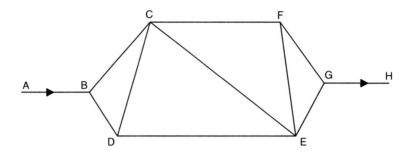

| Segment | Length (ft) | Diameter (in.) | Friction Factor |
|---------|-------------|----------------|-----------------|
| AB | 250 | 8 | 0.020 |
| BC | 450 | 12 | 0.020 |
| BD | 160 | 8 | 0.025 |
| CD | 500 | 12 | 0.030 |
| DE | 800 | 12 | 0.030 |
| CF | 600 | 12 | 0.020 |
| FE | 500 | 12 | 0.025 |
| CE | 800 | 12 | 0.030 |
| FG | 500 | 12 | 0.020 |
| EG | 200 | 8 | 0.030 |
| GH | 300 | 8 | 0.025 |

If the flow rate (incident at A) changes to 500 gal/min, the pressure loss (lb/in.$^2$) between nodes A and H is most nearly:

A. 65

B. 100

C. 50

D. 80

# 419

A 700,000-gal elevated water tank is used to provide fire fighting water demand for a school complex. The total effective area of the school is 160,000 ft$^2$. The school buildings are constructed of masonry. The time of fire protection (hours) that the tank can provide is most nearly:

A. 1.0
B. 1.5
C. 2.0
D. 2.5

## 420

A 280-acre development is under construction, which is estimated to last 3 years. During construction, the sediment yield is 1,200 ft³/acre-year. Following the completion of construction, the sediment yield is estimated to be 300 ft³/acre-year. A 5-acre water quality pond is used for sediment control. The bottom elevation of the pond is 345.6 ft above sea level and pool surface elevation is regulated at 353.4 ft above sea level. The minimum depth of water in the pond (that triggers sediment removal) is 3 ft. The number of years before the sediment must be removed from the pond is most nearly:

A. 1.8
B. 3.5
C. 4.3
D. 4.8

## 421

A deflector is used to deflect horizontal flow from a 2-in.-diameter nozzle to an angle of 40° to the horizontal. The flow rate through the nozzle is 500 gal/min. The pressure at a section immediately upstream of the nozzle is 60 psi. The force acting on the deflector (lb) is most nearly:

A. 25
B. 35
C. 75
D. 100

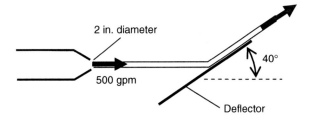

2 in. diameter

40°

500 gpm

Deflector

## 422

A stream has the following characteristics:

  Flow rate = 18 ft³/sec
  Average velocity = 4 ft/sec
  Temperature = 12°C
  $BOD_5$ = 2.0 mg/L
  Deoxygenation rate constant (log 10 at 20°C) = 0.20 day$^{-1}$
  Reoxygenation rate constant (log 10 at 20°C) = 0.30 day$^{-1}$
  D.O. = 5.1 mg/L

A factory discharges a wastewater stream into the river at point A. The wastewater has the following characteristics:

  Flow rate = 750 gal/min
  Temperature = 37°C
  $BOD_5$ = 105 mg/L
  D.O. = 1.7 mg/L

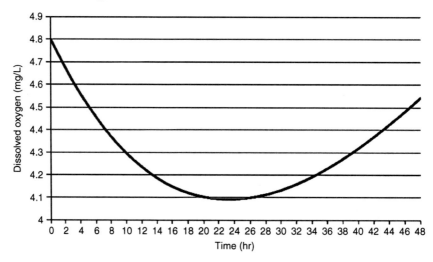

The dissolved oxygen sag curve for the stream-wastewater mix is shown in the curve above. The instant of mixing is considered $t = 0$. The oxygen deficit (mg/L) in the stream at a distance 20 mi downstream from point of mixing is most nearly:

A.  4.4

B.  6.4

C.  4.8

D.  4.1

## 423

A trapezoidal open channel with bottom width of 10 ft and side slopes 2H:1V conveys a flow rate of 150 ft³/sec. If the Manning's $n = 0.016$ and the bottom slope is 0.4%, the depth of flow (in.) is most nearly:

A.  15
B.  33
C.  29
D.  20

## 424

Water flows through a rectangular open channel at a constant flow rate. At a particular location of the channel, the floor of the channel is raised 4 in. for a 12-in.-long step, as shown in the figure below. The depth over the step is measured as 9 in. The flow rate per unit width of the channel (cfs/ft) is most nearly:

A.  1.57
B.  1.72
C.  1.85
D.  1.98

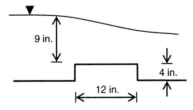

## 425

A channel and its floodplain (800 ft wide) have a 100-year flood water elevation of 252.7 ft above sea level. The mean ground elevation of the floodplain is 250.95 ft. Development within the floodplain encroaches to within 300 ft of the edge of the channel. Development in the 500-ft-wide area creates near impervious conditions.

(*Continued on next page*)

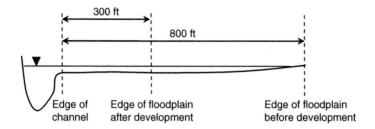

If the floodplain conveys 98% of the flood flow, what is most nearly the new 100-year flood elevation after development?

A. 252
B. 254
C. 256
D. 258

## 426

The schematic of an activated sludge process is shown below. The influent flow rate is 4 MGD. The primary clarifier removes 65% of the total suspended solids and 20% of the $BOD_5$. The primary sludge contains 6% solids. The quantity of primary sludge (gal/day) is most nearly:

A. 7,000
B. 9,000
C. 11,000
D. 14,000

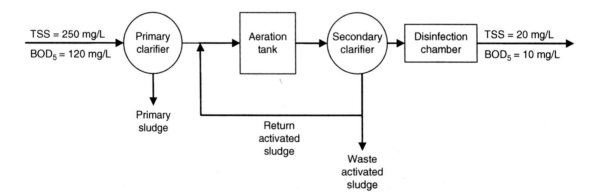

## 427

A water sample yields the following results:

| | |
|---|---|
| $Ca^{++}$ | 60.0 mg/L |
| $Mg^{++}$ | 21.2 mg/L |
| $Fe^{++}$ | 2.2 mg/L |
| $HCO_3^-$ | 221.3 mg/L |
| $Cl^-$ | 33.5 mg/L |

The hardness of the water sample (mg/L as $CaCO_3$) is most nearly:

A. 240

B. 243

C. 510

D. 230

## 428

A rectangular open channel with a width of 10 ft has a critical velocity of 15 ft/sec. The flow rate (ft³/sec) at this critical velocity is most nearly:

A. 800

B. 1,000

C. 1,200

D. 1,400

## 429

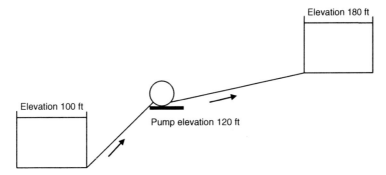

In the above diagram, the pump located at an elevation of 120 ft is used to pump a flow $Q = 3,000$ gal/min from a reservoir with surface elevation of 100 ft to another

(*Continued on next page*)

reservoir with surface elevation of 180 ft. The approximate efficiency of the pump is 85%. The characteristics of the pipe system are given below:

| | |
|---|---|
| Suction line: | 800-ft length, 18-in. diameter, C = 100 |
| Discharge line: | 2,500-ft length, 12-in. diameter, C = 100 |

The brake horsepower of the pump is most nearly:

A.  65

B.  75

C.  125

D.  150

## 430

A rectangular channel is 15 ft wide and conveys a flow rate of 120,000 gal/min. The flow is conveyed down a spillway of the same width, at the bottom of which the depth of flow is 15 in. If a hydraulic pump is to be forced at the bottom of the spillway, the tailwater depth (ft) needs to be most nearly:

A.  2.0

B.  3.4

C.  5.2

D.  6.2

## 431

A watershed (area = 370 acres) is subdivided into five distinct land use classifications, as shown in the following table. Storm runoff data have been abstracted into a set of intensity-duration-frequency curves. Using the NRCS method, the net runoff (in.) from a 20-year storm with gross rainfall of 5.6 in. is most nearly:

A.  3.6

B.  3.1

C.  1.5

D.  2.2

| Region | Area (acres) | Land Use | Soil Type | Time for Overland Flow (min) | Curve Number | Rational Runoff Coefficient |
|--------|--------------|----------|-----------|------------------------------|--------------|-----------------------------|
| A | 80 | Lawns: fair condition | B | 30 | 69 | 0.4 |
| B | 80 | Forest | C | 45 | 45 | 0.2 |
| C | 50 | Paved | B | 15 | 98 | 0.9 |
| D | 90 | Residential: 4 lots/acre | D | 25 | 87 | 0.6 |
| E | 70 | Forest | A | 45 | 35 | 0.2 |

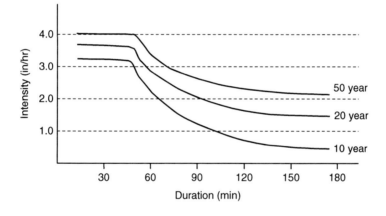

## 432

A municipal plant receives water with a total hardness of 200 mg/L. The designed discharge hardness is 50 mg/L. An ion exchange unit with an overall efficiency of 88% is used for hardness reduction. The bypass factor (%) is most nearly:

A. 10
B. 15
C. 20
D. 25

## 433

The parking lot shown below is located in Reno, Nevada (Region 7). The concrete pavement slopes toward the 400-ft-long gutter in the center of the lot. Rain, falling on the parking lot, drains into the gutter. The gutter-flow drains into a storm sewer inlet at one end. Only runoff from the parking lot enters this storm sewer inlet. Adjacent land drains elsewhere. For a significant rainfall event, the mean sheet flow velocity across the concrete parking lot is estimated to be 0.5 ft/sec. The estimated mean flow velocity in the gutter is 2.0 ft/sec. The 50-year recurrence interval peak discharge inflow (ft³/sec) to the storm sewer inlet is most nearly:

A. 13
B. 18
C. 23
D. 28

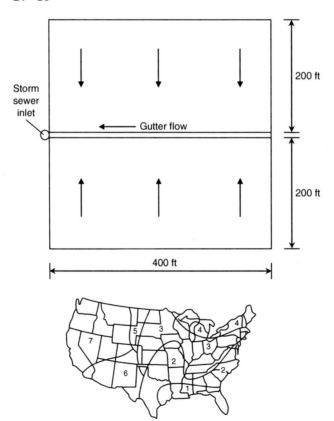

| Return Period (y) | Coefficients | Region | | | | | | |
|---|---|---|---|---|---|---|---|---|
| | | 1 | 2 | 3 | 4 | 5 | 6 | 7 |
| 2 | K | 206 | 140 | 106 | 70 | 70 | 68 | 32 |
| | b | 30 | 21 | 17 | 13 | 16 | 14 | 11 |
| 5 | K | 247 | 190 | 131 | 97 | 81 | 75 | 48 |
| | b | 29 | 25 | 19 | 16 | 13 | 12 | 12 |
| 10 | K | 300 | 230 | 170 | 111 | 111 | 122 | 60 |
| | b | 36 | 29 | 23 | 16 | 17 | 23 | 13 |
| 25 | K | 327 | 260 | 230 | 170 | 130 | 155 | 67 |
| | b | 33 | 32 | 30 | 27 | 17 | 26 | 10 |
| 50 | K | 315 | 350 | 250 | 187 | 187 | 160 | 65 |
| | b | 28 | 38 | 27 | 24 | 25 | 21 | 8 |
| 100 | K | 367 | 375 | 290 | 220 | 240 | 210 | 77 |
| | b | 33 | 36 | 31 | 28 | 29 | 26 | 10 |

## 434

For a community, it is estimated that a population's water consumption will double over the next 20 years. The cost of expanding the existing water supply system will be compared to a phased program of expansion. Immediate development would cost $420,000 with annual maintenance costs of $40,000. A phased program would involve an initial investment of $200,000 and an estimated expenditure of $320,000 in 10 years. Annual maintenance cost under the phased program is estimated to be $20,000 for the first 10 years, and $16,000 following that. Assume a *perpetual period of service* for each system and MARR = 7%. The ratio of the cost for phased program relative to the single investment program is most nearly:

A. 0.63
B. 0.95
C. 1.27
D. 1.62

## 435

A municipal wastewater treatment plant treats a flow rate of 3 MGD containing $BOD_5 = 240$ mg/L. BOD removal takes place in a single-stage rock media trickling filter. Filter dimensions are: diameter = 80 ft, depth = 6 ft. Recirculation rate $R = 3$. The $BOD_5$ of the effluent is most nearly:

A. 60
B. 80
C. 120
D. 180

## 436

A 30-ft-deep aquifer is confined between two impermeable layers of rock, as shown below. The elevation of the piezometric surface is at 20 ft above the top of the aquifer. A 9-in.-diameter well is used to establish a steady state pumping rate of 2,000 gal/min. The hydraulic conductivity of the soil in the aquifer is 1,000 ft/day. Observation wells 1 and 2 are drilled at radial distances of 30 ft and 180 ft, respectively, from the centerline of the pumping well, as shown. The drawdown of the piezometric surface at observation well no. 1 is 4.5 ft. The drawdown of the piezometric surface (ft) at observation well no. 2 is most nearly:

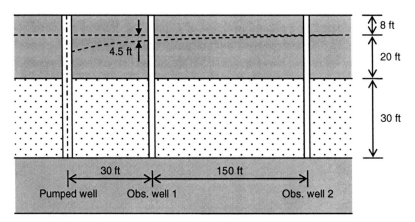

A. 2.1
B. 2.8
C. 3.2
D. 3.8

## 437

An anaerobic digester has a VSS removal efficiency of 60%. The digester is used to treat a wastewater flow of 3 MGD containing suspended solids concentration of 450 mg/L. The suspended solids are 65% volatile. The total suspended solids load (lb/day) in the effluent is most nearly:

A. 5,000
B. 6,000
C. 7,000
D. 8,000

## 438

Results for a settling column analysis are summarized in the form of percent removal curves as shown below. The cumulative removal rate at a depth of 3 ft at $t = 90$ min is most nearly:

A. 43
B. 57
C. 65
D. 75

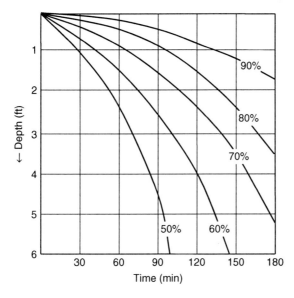

## 439

A secondary effluent at a wastewater treatment plant flows at a rate of 3.5 MGD. The effluent has the following characteristics: $BOD_5 = 20$ mg/L, TSS = 35 mg/L, pH = 7.5, temp = 15°C. The secondary effluent is disinfected by chlorination with a desired 2-log inactivation. Chlorine demand of the secondary effluent has been estimated as 3.5 mg/L. If the chlorine dose is 5.0 mg/L, the minimum required volume of the chlorination chamber ($ft^3$) is most nearly:

A. 14,700

B. 17,500

C. 20,200

D. 23,100

### CT Values for Various Levels of Disinfection

| Chlorine Residual Conc. (mg/L) | CT (mg/L-min) for Log-Inactivation | | | | |
|---|---|---|---|---|---|
| | 1.0 | 1.5 | 2.0 | 2.5 | 3.0 |
| 0.5 | 29 | 43 | 58 | 72 | 86 |
| 1.0 | 16 | 48 | 64 | 80 | 96 |
| 1.5 | 17 | 51 | 68 | 85 | 102 |
| 2.0 | 18 | 53 | 71 | 89 | 106 |
| 2.5 | 18 | 55 | 73 | 92 | 110 |
| 3.0 | 19 | 56 | 75 | 94 | 113 |
| 3.5 | 19 | 58 | 77 | 96 | 116 |

## 440

A bank of anaerobic digesters, arranged in parallel, is used to treat a wastewater flow of 4 MGD containing TSS = 540 mg/L (75% volatile). The following design parameters of a digester are given:

Minimum hydraulic detention time = 2 hr

Maximum organic load = $2.1 \times 10^{-3}$ lb-VSS/gal-hr

Maximum diameter = 6 ft

Which of the following arrangements is adequate?
A.   12 units, each with diameter = 6 ft and length = 100 ft
B.   16 units, each with diameter = 4 ft and length = 120 ft
C.   12 units, each with diameter = 6 ft and length = 150 ft
D.   16 units, each with diameter = 4 ft and length = 200 ft

## END OF WATER RESOURCES & ENVIRONMENTAL DEPTH EXAM

## USE THE ANSWER SHEET ON THE NEXT PAGE

# Water Resources & Environmental Depth Exam: Answer Sheet

| | | | | |
|---|---|---|---|---|
| 401 | Ⓐ | Ⓑ | Ⓒ | Ⓓ |
| 402 | Ⓐ | Ⓑ | Ⓒ | Ⓓ |
| 403 | Ⓐ | Ⓑ | Ⓒ | Ⓓ |
| 404 | Ⓐ | Ⓑ | Ⓒ | Ⓓ |
| 405 | Ⓐ | Ⓑ | Ⓒ | Ⓓ |
| 406 | Ⓐ | Ⓑ | Ⓒ | Ⓓ |
| 407 | Ⓐ | Ⓑ | Ⓒ | Ⓓ |
| 408 | Ⓐ | Ⓑ | Ⓒ | Ⓓ |
| 409 | Ⓐ | Ⓑ | Ⓒ | Ⓓ |
| 410 | Ⓐ | Ⓑ | Ⓒ | Ⓓ |
| 411 | Ⓐ | Ⓑ | Ⓒ | Ⓓ |
| 412 | Ⓐ | Ⓑ | Ⓒ | Ⓓ |
| 413 | Ⓐ | Ⓑ | Ⓒ | Ⓓ |
| 414 | Ⓐ | Ⓑ | Ⓒ | Ⓓ |
| 415 | Ⓐ | Ⓑ | Ⓒ | Ⓓ |
| 416 | Ⓐ | Ⓑ | Ⓒ | Ⓓ |
| 417 | Ⓐ | Ⓑ | Ⓒ | Ⓓ |
| 418 | Ⓐ | Ⓑ | Ⓒ | Ⓓ |
| 419 | Ⓐ | Ⓑ | Ⓒ | Ⓓ |
| 420 | Ⓐ | Ⓑ | Ⓒ | Ⓓ |

| | | | | |
|---|---|---|---|---|
| 421 | Ⓐ | Ⓑ | Ⓒ | Ⓓ |
| 422 | Ⓐ | Ⓑ | Ⓒ | Ⓓ |
| 423 | Ⓐ | Ⓑ | Ⓒ | Ⓓ |
| 424 | Ⓐ | Ⓑ | Ⓒ | Ⓓ |
| 425 | Ⓐ | Ⓑ | Ⓒ | Ⓓ |
| 426 | Ⓐ | Ⓑ | Ⓒ | Ⓓ |
| 427 | Ⓐ | Ⓑ | Ⓒ | Ⓓ |
| 428 | Ⓐ | Ⓑ | Ⓒ | Ⓓ |
| 429 | Ⓐ | Ⓑ | Ⓒ | Ⓓ |
| 430 | Ⓐ | Ⓑ | Ⓒ | Ⓓ |
| 431 | Ⓐ | Ⓑ | Ⓒ | Ⓓ |
| 432 | Ⓐ | Ⓑ | Ⓒ | Ⓓ |
| 433 | Ⓐ | Ⓑ | Ⓒ | Ⓓ |
| 434 | Ⓐ | Ⓑ | Ⓒ | Ⓓ |
| 435 | Ⓐ | Ⓑ | Ⓒ | Ⓓ |
| 436 | Ⓐ | Ⓑ | Ⓒ | Ⓓ |
| 437 | Ⓐ | Ⓑ | Ⓒ | Ⓓ |
| 438 | Ⓐ | Ⓑ | Ⓒ | Ⓓ |
| 439 | Ⓐ | Ⓑ | Ⓒ | Ⓓ |
| 440 | Ⓐ | Ⓑ | Ⓒ | Ⓓ |

# 6

# Transportation Depth Exam

The following set of questions numbered 501 to 540 is representative of a 4-hour transportation depth exam according to the syllabus and guidelines for the Principles and Practice (P&P) of Civil Engineering Examination administered by the National Council of Examiners for Engineering and Surveying (NCEES). Detailed solutions are on page 293.

## 501

A horizontal compound curve connects two tangents as shown below. The bearing of the back tangent is N46°25'32" E, and the bearing of the forward tangent is S17°56'21" E. The curve radii to be used are $R_1 = 1,800$ ft and $R_2 = 900$ ft. The PC is located at station $138 + 34.12$ and the PI at station $158 + 12.98$.

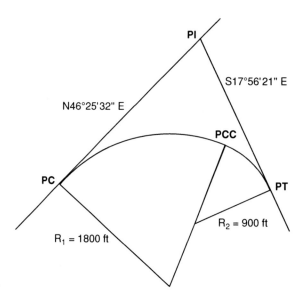

The station of the PCC is most nearly:

A. $145 + 12.54$
B. $147 + 05.23$
C. $148 + 50.61$
D. $151 + 03.32$

## 502

A parabolic crest curve is followed by a sag curve. The two curves are connected by a tangent section, as shown. A bridge structure is located at station $45 + 55.00$. The elevation of the low point on the bridge is 405.54 ft. The design speed (mph) of a truck (assume driver's eye elevation = 7.6 ft), based on adequate sight distance under the bridge overpass on curve number 2, is most nearly:

A. 65
B. 75
C. 80
D. 85

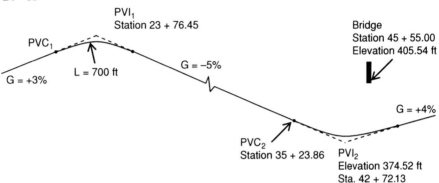

## 503

A circular horizontal curve connects two tangents with a deflection angle of 42°50'. The roadway is a two-lane highway with 12-ft lanes. The centerline of the roadway has a radius of 1,245 ft. The design speed = 50 mph. PC for the centerline circular curve is at sta. 23 + 32.56. It is proposed to insert a pair of symmetric spirals at the PC and the PT. The minimum length of spiral (ft) is most nearly:

A. 120
B. 140
C. 160
D. 180

## 504

A 6-lane freeway through rolling terrain (rural) has12-ft lanes, interchanges every 1.25 mi and has 8-ft-wide shoulders. Traffic studies have resulted in the following data:

AADT = 65,000 veh/day.
The traffic stream includes 8% trucks, 3% buses and 5% RVs.
K = 0.12
Peak hour factor, based on 15-min traffic counts, is 0.9.
During peak flow, directional split = 60/40.

*(Continued on next page)*

The LOS for the peak direction of the freeway is most nearly:

A. A
B. B
C. C
D. D

# 505

A study area is divided into three socio-economic zones, whose trip production and trip attractions are tabulated below:

| Zone | 1 | 2 | 3 | Total |
|---|---|---|---|---|
| Trip Productions | 250 | 450 | 440 | 1,140 |
| Trip Attractions | 120 | 350 | 670 | 1,140 |

Resistance to travel between zones is represented by the following matrix of friction factors, which are approximately inversely proportional to time of travel between zones.

| Zone | 1 | 2 | 3 |
|---|---|---|---|
| 1 | 40 | 90 | 75 |
| 2 | 90 | 25 | 35 |
| 3 | 75 | 35 | 40 |

The number of trips produced by zone 3 and attracted to zone 2, according to the first iteration of the gravity model, is most nearly:

A. 110
B. 130
C. 150
D. 170

# 506

A 4-lane divided highway (2 lanes in each direction) currently has ADT of 23,000 vehicles (each way). Approximately 12% of the traffic consists of trucks. Traffic is expected to grow by a 4% annual rate over the next 20 years. Table 1 shows data collected from a weigh station during a typical day for 1,078 trucks.

| Table 1: Traffic Data (Axle Counts) | | |
| --- | --- | --- |
| Axle Load (k) | Number of Single Axles | Number of Tandem Axles |
| 10 | 2420 | |
| 14 | 630 | |
| 18 | 301 | |
| 20 | 22 | |
| 22 | 6 | 24 |
| 25 | 1 | 15 |
| 28 | | 12 |
| 32 | | 11 |

| Table 2: Load Equivalence Factors | |
| --- | --- |
| Single Axles | |
| Axle Load (k) | Load Equivalence Factor |
| 10 | 0.0877 |
| 14 | 0.360 |
| 18 | 1.0 |
| 20 | 1.51 |
| 22 | 2.18 |
| 25 | 3.53 |
| Tandem Axles | |
| Axle Load (k) | Load Equivalence Factor |
| 22 | 0.180 |
| 25 | 0.308 |
| 28 | 0.495 |
| 32 | 0.857 |

Using the Asphalt Institute Load Equivalence Factors (LEF) shown in Table 2, the design 20-year 18-k ESAL ($W_{18}$) is most nearly

A.  $9 \times 10^6$
B.  $15 \times 10^6$
C.  $20 \times 10^6$
D.  $23 \times 10^6$

## 507

A 6-lane highway has the following characteristics:

Level terrain
Lane width = 11-ft lanes
Average spacing between accessing driveways = 600 ft
Right shoulder width = 6 ft
Clear distance to nearest obstruction in median = 4 ft

*(Continued on next page)*

Directional hourly volume = 3,440 veh/hr
The traffic stream includes 8% trucks, 3% buses, and 2% RVs.
Peak hour factor, based on 15-min traffic counts, is 0.88.
Drivers are mostly commuters.

The LOS for the peak direction of the highway is most nearly:

A. B
B. C
C. D
D. F

## 508

A transition curve is to be used to implement a change in cross-section from a normally crowned section to a fully superelevated section. The outer lane is to be rotated from the normally crowned section to a straight level section at the tangent-to-spiral (T.S.) station. The roadway section is rotated about the centerline. The following information is given:

Degree of curve (D) = 2.5°
Two 11-ft lanes
Design superelevation = 0.08 ft/ft
Grade = +1.50%
Crown = 0.015 ft/ft
P.I. station = 100 + 0.00
Deflection angle between tangents = 56°35'48"
Length of spiral = 230 ft

The station (ft) of the T.S. is most nearly:

A. 86 + 50
B. 87 + 25
C. 88 + 00
D. 88 + 50

## 509

What is the recommended length of cycle for a 4-phase signal with 3-sec lost time per phase, with the following critical movements?

| Phase | Lane Group Volume (vph) | Saturation Volume for Lane Group (vphg) | Green Time Fraction |
|-------|-------------------------|------------------------------------------|---------------------|
| 1 | 100 | 900 | 0.11 |
| 2 | 600 | 1,900 | 0.36 |
| 3 | 105 | 900 | 0.13 |
| 4 | 630 | 2,300 | 0.40 |

A. 105 sec
B. 125 sec
C. 140 sec
D. 160 sec

*Note:* The following diagram and the associated data are to be used for problems 510–512.

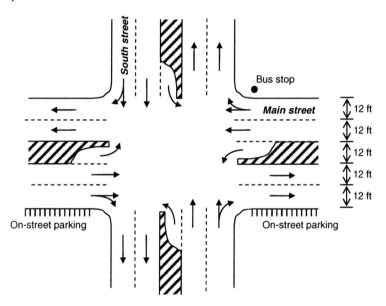

A right-angle intersection between two 4-lane highways (Main St. and South St.) is shown above. The 80-sec signal cycle has four phases: (1) EBL and WBL, (2) EBTH/R and WBTH/R, (3) NBL and SBL, and (4) NBTH/R and SBTH/R.

*(Continued on next page)*

Each yellow time separating phase is 3 sec and each all-read interval between phases is 1 sec. The critical v/s ratios for each phase are listed below:

| Phase | Critical Movement | v/s |
|-------|-------------------|-----|
| 1 | WBL | 0.232 |
| 2 | EBTH/R | 0.156 |
| 3 | NBL | 0.150 |
| 4 | SBTH/R | 0.135 |

Posted speed on both approaches = 40 mph
Heavy vehicles = 5%
PHF = 0.92
25% of TH/R traffic is RT
10 buses/hr on EBTH/R and WBTH/R approaches
25 parking maneuvers per hour for EBTH/R and WBTH/R approaches
Perception-reaction time = 1 sec
Deceleration rate = 10 ft/sec$^2$
Vehicle length = 20 ft

Pedestrian volumes:      On north-south walkways (10 ft wide) = 1,200 pedestrians/hr
On east-west walkways (10 ft wide) = 800 pedestrians/hr

## 510

The saturation flow (vph) for the WBTH/R lane group is most nearly:

A. 2,650
B. 2,860
C. 3,030
D. 3,210

## 511

(See figure and data associated with problem 510)

The minimum clearance interval (sec) for the intersection is most nearly:

A. 4.0
B. 4.5
C. 5.3
D. 5.8

## 512

(See figure and data associated with problem 510)

The minimum green time for the north-south signal phase, based on pedestrian volume, is most nearly:

A. 25 sec
B. 22 sec
C. 19 sec
D. 16 sec

## 513

A circular horizontal curve has PC at 12 + 43.56. Degree of curve = 4°. Deflection angle between tangents is 56°24'45" (right). In order to provide adequate horizontal sightline offset at the design speed, it is proposed to shift the forward tangent by a parallel offset of 20 ft, while maintaining the PC, as shown in the figure. The new degree of curve is most nearly:

A. 3.7
B. 3.8
C. 3.9
D. 4.1

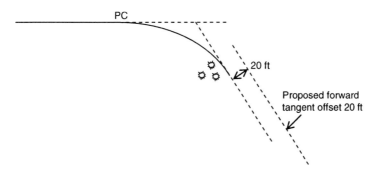

## 514

An agency is investigating installing a signal at an intersection in a large city where the 85th percentile speeds are 40 mph on both approaches. The intersection

(*Continued on next page*)

has two lanes in the major direction and one lane in the minor direction. Recent traffic counts from the high-volume hours at the intersection are provided below:

| Time Period | Vehicles on Both Major Approaches Combined | Vehicles on Highest Volume Minor Street Approach |
|---|---|---|
| 7:00–8:00 A.M. | 600 | 210 |
| 8:00–9:00 A.M. | 650 | 230 |
| 11:00–12:00 A.M. | 550 | 210 |
| 12:00–1:00 P.M. | 730 | 250 |
| 3:00–4:00 P.M. | 660 | 260 |
| 4:00–5:00 P.M. | 830 | 275 |
| 5:00–6:00 P.M. | 990 | 310 |
| 6:00–7:00 P.M. | 800 | 250 |

Which warrant(s) from the 2009 MUTCD does this intersection meet?
A. Warrants 1 & 2
B. Warrant 2 only
C. Warrants 1 & 3
D. Warrants 2 & 3

## 515

A reinforced-concrete cantilever retaining wall is shown below. The friction angle between the wall footing and the soil is 20°. The factor of safety for overturning of the wall is most nearly:

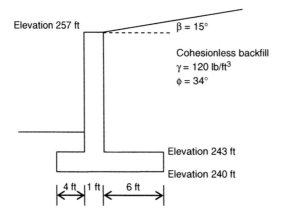

Elevation 257 ft

β = 15°

Cohesionless backfill
γ = 120 lb/ft³
φ = 34°

Elevation 243 ft
Elevation 240 ft

4 ft  1 ft  6 ft

A. 1.2
B. 2.5
C. 3.4
D. 4.8

## 516

A 3-lane freeway interchange in a suburban area has a 1-lane diverging ramp. The adjusted peak-hour volume just downstream from the diverge of interest is 4,100 passenger cars, while the adjusted peak-hour volume on the off-ramp is 500 passenger cars. The deceleration lane is 1,100-ft long, the ramp free-flow speed is 54 mph, and the ramp has a +3% grade. An off ramp 1,250 ft downstream of the diverge of interest serves an adjusted peak hour volume of 450 passenger cars.

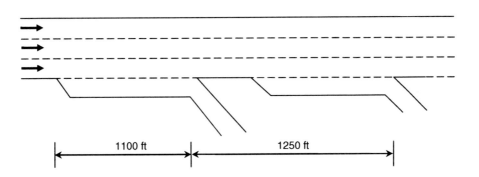

The adjusted traffic volume in the right two lanes of the freeway just upstream of the diverge of interest is most nearly:
A. 2,800
B. 2,900
C. 2,950
D. 3,050

## 517

A traffic network within a city planning zone is described by the following matrix. Nodes are numbered 1 through 8.

| Node | 1 | 2 | 3 | 4 | 5 | 6 | 7 | 8 |
|------|---|---|---|---|---|---|---|---|
| 1 | – | 1 | 0 | 1 | 0 | 1 | 0 | 1 |
| 2 | 1 | – | 1 | −1 | 0 | 0 | 0 | 0 |
| 3 | 0 | 1 | – | 1 | 1 | 0 | 0 | 0 |
| 4 | 1 | 1 | 1 | – | 1 | 1 | 1 | 0 |
| 5 | 0 | 0 | 1 | 1 | – | 1 | 0 | 0 |
| 6 | 1 | 0 | 0 | 1 | −1 | – | 1 | 0 |
| 7 | 0 | 0 | 0 | 1 | 0 | 1 | – | 1 |
| 8 | 1 | 0 | 0 | 0 | 0 | 0 | 1 | – |

The number of links in the network is most nearly:

A. 14
B. 15
C. 16
D. 17

## 518

A 2-lane rural freeway segment has 12-ft lanes and an 8-ft shoulder. A cut slope of 1V:4H exists adjacent to the shoulder as shown. The ADT is 16,000 and the design speed is 65 mph. The minimum width (ft) measured from the edge of the shoulder, which should be free of any obstructions, is most nearly:

A. 18
B. 26
C. 38
D. 46

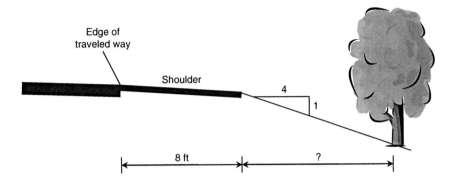

## 519

A 2-lane Class I highway in rolling terrain has the following characteristics:

Lane width = 12 ft, shoulder width = 6 ft
40% no passing zones
2-directional flow rate = 2,800 pc/hr
Directional split = 70/30
10% trucks and buses, 6% RVs
PHF = 0.85

For the design direction of the highway, the average travel speed has been calculated as 42 mph and the percent-time spent following has been calculated as 63%.

The level of service for the design direction is:

A. B
B. C
C. D
D. F

## 520

A horizontal circular curve is to connect a back-tangent bearing S42°30'W to a forward tangent bearing N70°W. If the degree of curve is 3°45' and the tangents intersect at station 50 + 22.30, the deflection angle for station 57 + 00.00 is most nearly:

A. 31°50'56"
B. 21°40'28"
C. 41°32'15"
D. 27°40'24"

## 521

A parabolic vertical curve joins a grade of −5% to a grade of +3%. The PVC is at station 53 + 12.50 and the PVI is at station 60 + 09.00. Elevation of the PVI is 365.57 ft. The curve passes under a bridge structure at station 55 + 05.20. The bottom elevation of the bridge is 405.20 ft. The vertical clearance under the bridge (ft) is most nearly:

A. 12.5

B. 13.4

C. 14.2

D. 14.9

## 522

A 2-lane highway has a circular horizontal alignment with centerline radius of 750 ft. Lanes are 12 ft wide. An obstructing structure exists 10 ft from the inside edge of the roadway as shown below. The safe design speed (mph), based on adequate horizontal sightline offset, on the curve is most nearly:

A. 35

B. 40

C. 45

D. 50

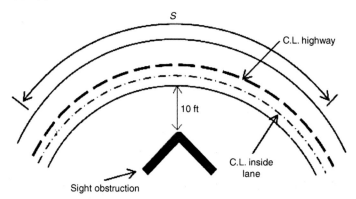

## 523

A 2-lane highway has a posted speed limit of 50 mph. The alignment consists of a dangerous curve where the maximum posted speed is 20 mph. What is the distance (ft) upstream of the low speed limit section where a warning sign must be posted?

A. 125
B. 200
C. 150
D. 175

## 524

A horizontal curve has a radius of 1,150 ft as shown below. A point P on the major chord is located such that AP = 215 ft. The length of the chord offset PQ at point P (ft) is most nearly:

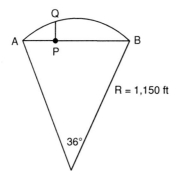

A. 43
B. 48
C. 52
D. 21

## 525

A vertical curve is being designed to connect two grades with $G_1 = +2.1\%$ and $G_2 = -1.5\%$. If the design speed is 65 mph, the minimum length of vertical curve (ft) to satisfy AASHTO safe stopping distance criteria is most nearly:
A. 600
B. 650
C. 800
D. 700

## 526

A parking lot is open during the hours of 8 A.M. to 6 P.M. During this time, exactly 360 cars were parked on the lot; 10% for 1 hr, 15% for 2 hr, 20% for 3 hr, 30% for 4 hr, and the remaining for the entire day. On average, 15% of the spaces are vacant and the operational efficiency factor is 80%.

The space-hr demand and number of parking spaces in the lot are most nearly:
A. 1,990 space-hr and 250 spaces
B. 1,690 space-hr and 240 spaces
C. 1,690 space-hr and 170 spaces
D. 1,990 space-hr and 200 spaces

## 527

A 6-lane freeway through mountainous terrain has ADT of 52,500 veh/day. The traffic stream includes 8% trucks, 3% buses, and 5% RVs. The FFS is estimated as 70 mph and the PHF = 0.9. Directional split is 60/40 and K = 0.12.

The LOS for the peak direction of the road is most nearly:
A. B
B. C
C. D
D. E

## 528

A new highway with a design speed of 50 mph is to be connected to an existing highway by a pair of reverse 2° curves. The new highway is parallel to the existing highway, offset 80 ft to the right.

The length of the reverse curve transition (ft), from PC to PT, is most nearly:
A. 1,250
B. 480
C. 960
D. 630

## 529

A 3-legged intersection is shown below. The minor street approach to the intersection has two lanes, is controlled by a STOP sign, has a grade of +4% heading into the intersection, and serves 15% trucks during the peak hour. The major street has a basically flat grade, a speed limit of 45 mph, and serves 20% trucks during the peak hour. The median is 6 ft wide.

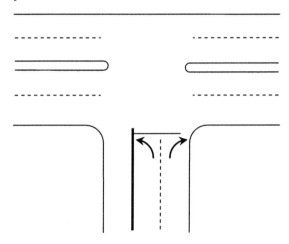

The critical gap for the left turn movement from the minor street during the peak hour is most nearly:

A. 7.1 sec
B. 7.5 sec
C. 7.8 sec
D. 8.5 sec

## 530

A traffic stream has mean headways of 2.4 sec. If the jam density is 64 veh/mi, the optimum speed is 50 mph, and the flow is 1,500 veh/hr, then the capacity of the highway (veh/hr) is most nearly:

A. 1,530
B. 1,600
C. 1,640
D. 1,720

## 531

The following data is given for a soil sample:

| | Sieve Analysis: | | Atterberg Tests: |
|---|---|---|---|
| Sieve size | Percent Retained | | Liquid Limit = 43 |
| No. 4 | 8 | | Plastic Limit = 21 |
| No. 10 | 10 | | |
| No. 20 | 12 | | |
| No. 40 | 21 | | |
| No. 100 | 15 | | |
| No. 200 | 8 | | |

The AASHTO soil classification is:

A.  A-6

B.  A-7

C.  A-2-6

D.  A-2-7

## 532

The table below shows recorded discharge at a stream monitoring station following a 2-hr storm. The tributary watershed area contributing runoff to the stream has been established as 115 acres.

| Time (hr) | 0 | 1 | 2 | 3 | 4 | 5 | 6 |
|---|---|---|---|---|---|---|---|
| Discharge Q (ft³/sec) | 23 | 84 | 127 | 112 | 75 | 32 | 25 |

The peak stream discharge (ft³/sec) that would be recorded following a 2-hr storm producing 1.7 in. of runoff is most nearly:

A.  90

B.  80

C.  180

D.  160

## 533

A 36-in.-diameter reinforced-concrete pipe under a roadway serves as a culvert operating under inlet control. The following data apply to the culvert:

Upstream invert: Elevation 325.0
Downstream invert: Elevation 322.5
Length of culvert: 145 ft
Entrance condition: Headwall with square edge
Roadway elevation: 337.8 ft
Culvert flow rate due to 100-year flood event: 100 cfs

The minimum vertical clearance (ft) between the roadway and the 100-year flood elevation is most nearly:

A. 5.2
B. 2.2
C. 4.8
D. 7.8

## 534

A trapezoidal open channel with bottom width of 10 ft and side slopes 2H:1V conveys a flow rate of 150 ft³/sec. If the Manning's $n$ is 0.016 and the bottom slope is 0.4%, the depth of flow (in.) is most nearly:

A. 12
B. 16
C. 20
D. 24

## 535

Results from a standard Proctor compaction test of six soil samples from a borrow pit are tabulated. The natural moisture content of the excavated material is 12%. The fill location requires 1.5 million yd³ of soil compacted to a minimum 90% of the maximum Proctor dry density.

(*Continued on next page*)

| Sample | Net Weight of Soil (lb) | Moisture Content (%) |
|--------|-------------------------|----------------------|
| 1 | 3.24 | 12 |
| 2 | 3.70 | 14 |
| 3 | 3.95 | 16 |
| 4 | 4.21 | 18 |
| 5 | 3.90 | 20 |
| 6 | 3.40 | 22 |

The total volume of borrow soil that must be excavated (yd$^3$) is most nearly:

A. 1.72 million

B. 1.65 million

C. 1.53 million

D. 1.42 million

## 536

A 4-leg intersection in an urban location had 15 recorded accidents (all types) during 2005. During this year, the average daily volumes entering the intersection from the four approaches were 1,900, 1,270, 1,620, and 930 vehicles. The intersection's accident rate (per hundred million entering vehicles) is most nearly:

A. 7.2

B. 72

C. 720

D. 7,200

## 537

A temporary road closure (not exceeding 20 min during the daytime) is established around a 1,250-ft-long work zone (including a buffer space) on a freeway. The distance of the ROAD WORK AHEAD sign (ft) from the centerline of the work zone is most nearly:

A. 4,800

B. 5,800

C. 6,800

D. 7,800

## 538

Ground-mounted signs are installed at the side of the road in a commercial district where pedestrian movement or parking is likely to occur. According to the MUTCD, the vertical distance (ft) between the bottom of the sign and the top of the near edge of the traveled path is most nearly:

A.  4
B.  5
C.  7
D.  8

## 539

The table below shows areas of cut and fill sections at locations 50 ft apart. Assume shrinkage = 12% and bulking = 25%. The cumulative earthwork volume between stations 0 + 0.00 and 3 + 0.00 is most nearly:

A.  1,400
B.  1,800
C.  2,200
D.  2,600

| Station | Area (ft²) | |
| --- | --- | --- |
|  | Cut | Fill |
| 0 + 00.00 | 563.2 | 342.2 |
| 0 + 50.00 | 213.5 | 213.6 |
| 1 + 00.00 | 123.5 | 343.3 |
| 1 + 50.00 | 654.6 | 111.0 |
| 2 + 00.00 | 973.1 | 762.4 |
| 2 + 50.00 | 567.3 | 342.9 |
| 3 + 00.00 | 451.6 | 190.4 |

## 540

In order to meet current demand, a 1-mi segment of an urban highway must be widened by two lanes. The cost of the improvement will be compared to a phased program of expansion. Immediate development would cost $4,200,000 with annual maintenance costs of $40,000. A phased program would involve an initial investment of $2,000,000 and an estimated expenditure of $3,200,000 in 10 years. Annual maintenance cost under the phased program is estimated to be $28,000 for the first 10 years and $36,000 following that. Assume a *perpetual period of service* for each system and MARR of 7%. The ratio of the cost for a phased program relative to the single investment program is most nearly:

A. 0.85

B. 0.95

C. 1.04

D. 1.17

**END OF TRANSPORTATION DEPTH EXAM**

**USE THE ANSWER SHEET ON THE NEXT PAGE**

# Transportation Depth Exam: Answer Sheet

| | | | | | | | | | | |
|---|---|---|---|---|---|---|---|---|---|---|
| 501 | Ⓐ | Ⓑ | Ⓒ | Ⓓ | | 521 | Ⓐ | Ⓑ | Ⓒ | Ⓓ |
| 502 | Ⓐ | Ⓑ | Ⓒ | Ⓓ | | 522 | Ⓐ | Ⓑ | Ⓒ | Ⓓ |
| 503 | Ⓐ | Ⓑ | Ⓒ | Ⓓ | | 523 | Ⓐ | Ⓑ | Ⓒ | Ⓓ |
| 504 | Ⓐ | Ⓑ | Ⓒ | Ⓓ | | 524 | Ⓐ | Ⓑ | Ⓒ | Ⓓ |
| 505 | Ⓐ | Ⓑ | Ⓒ | Ⓓ | | 525 | Ⓐ | Ⓑ | Ⓒ | Ⓓ |
| 506 | Ⓐ | Ⓑ | Ⓒ | Ⓓ | | 526 | Ⓐ | Ⓑ | Ⓒ | Ⓓ |
| 507 | Ⓐ | Ⓑ | Ⓒ | Ⓓ | | 527 | Ⓐ | Ⓑ | Ⓒ | Ⓓ |
| 508 | Ⓐ | Ⓑ | Ⓒ | Ⓓ | | 528 | Ⓐ | Ⓑ | Ⓒ | Ⓓ |
| 509 | Ⓐ | Ⓑ | Ⓒ | Ⓓ | | 529 | Ⓐ | Ⓑ | Ⓒ | Ⓓ |
| 510 | Ⓐ | Ⓑ | Ⓒ | Ⓓ | | 530 | Ⓐ | Ⓑ | Ⓒ | Ⓓ |
| 511 | Ⓐ | Ⓑ | Ⓒ | Ⓓ | | 531 | Ⓐ | Ⓑ | Ⓒ | Ⓓ |
| 512 | Ⓐ | Ⓑ | Ⓒ | Ⓓ | | 532 | Ⓐ | Ⓑ | Ⓒ | Ⓓ |
| 513 | Ⓐ | Ⓑ | Ⓒ | Ⓓ | | 533 | Ⓐ | Ⓑ | Ⓒ | Ⓓ |
| 514 | Ⓐ | Ⓑ | Ⓒ | Ⓓ | | 534 | Ⓐ | Ⓑ | Ⓒ | Ⓓ |
| 515 | Ⓐ | Ⓑ | Ⓒ | Ⓓ | | 535 | Ⓐ | Ⓑ | Ⓒ | Ⓓ |
| 516 | Ⓐ | Ⓑ | Ⓒ | Ⓓ | | 536 | Ⓐ | Ⓑ | Ⓒ | Ⓓ |
| 517 | Ⓐ | Ⓑ | Ⓒ | Ⓓ | | 537 | Ⓐ | Ⓑ | Ⓒ | Ⓓ |
| 518 | Ⓐ | Ⓑ | Ⓒ | Ⓓ | | 538 | Ⓐ | Ⓑ | Ⓒ | Ⓓ |
| 519 | Ⓐ | Ⓑ | Ⓒ | Ⓓ | | 539 | Ⓐ | Ⓑ | Ⓒ | Ⓓ |
| 520 | Ⓐ | Ⓑ | Ⓒ | Ⓓ | | 540 | Ⓐ | Ⓑ | Ⓒ | Ⓓ |

# 7

# Construction Depth Exam

The following set of questions numbered 601 to 640 is representative of a 4-hour construction depth exam according to the syllabus and guidelines for the Principles and Practice (P&P) of Civil Engineering Examination administered by the National Council of Examiners for Engineering and Surveying (NCEES). Detailed solutions are on page 315.

## 601

An excavator has a bucket capacity of 2.8 yd³. Its operation cycle consists of the following phases: (a) excavation time = 45 sec, (b) travel time (two-way) = 4 min, and (c) dumping/transfer time = 30 sec. Assume an overall efficiency factor for the excavator = 85%.

The quantity of material to be excavated (bank measure) = 50,000 ft³. The material has the following properties:

Swell = 20%
*In situ* unit weight = 120 lb/ft³
Water content = 30%

Assuming 8-hr workdays, the number of days required to complete the job is most nearly:

A. 9
B. 10
C. 11
D. 13

## 602

An excavator has a bucket capacity of 3.0 yd³. Its operation cycle consists of the following phases: (a) excavation time = 45 sec, (b) travel time (two-way) = 4 min, and (c) transfer time (to trucks) = 45 sec. The excavator transfers the excavated material to a fleet of trucks that carry the material offsite. The following data is given:

Overall efficiency factor for the excavator = 90%
Truck capacity = 15 yd³
Truck cycle time (transfer + two-way travel + dumping) = 90 min
The quantity of excavated material = 2,600 yd³ (loose)

The number of trucks needed to balance the production of the excavator, assuming 8-hr workdays, is most nearly:

A. 2
B. 3
C. 4
D. 5

## 603

The architect for a warehouse building has proposed a design change to ensure compatibility with local building codes. The revised SOW is to substitute one layer of 5/8-in.-thick sheet rock with two layers of ½-in.-thick fire code (FC) GWB and to provide insulation within the wall cavity.

The building plan dimensions are: 180 ft × 200 ft
Floor-to-ceiling height: 12 ft
Openings: eight 6 ft 0 in. wide × 10 ft 0 in. high openings

### Labor rates

| | |
|---|---|
| Carpenter foreman (working) | $50.00 fully burdened |
| Carpenter (journeyman) | $40.00 fully burdened |
| Laborer | $25.00 fully burdened |

### Work crew

4 carpenters
2 laborers
1 working foreman

### Work crew productivity (based on 8 hr/day)

| | |
|---|---|
| GWB installation | 960 ft²/L.H. |
| Insulation | 1,920 ft²/L.H. |

### Material costs

(add 10% waste factor for all materials)

| | |
|---|---|
| 4 ft 0 in. × 10 ft 0 in. × ½ in. GWB (FC) | $0.285/ft² |
| 4 ft 0 in. × 10 ft 0 in. × 5/8 in. GWB | $0.255/ft² |
| Insulation | $0.45/ft² |
| Contractor's overhead | 10% |
| Contractor's profit | 5% |

The complete cost ($) of the change order for the revised SOW is most nearly:

A. 3,300
B. 4,200
C. 9,800
D. 7,800

## 604

A 450-ft-long canal (trapezoidal section: 19-ft-bottom width, 3H:2V side slopes) is to be lined with concrete to a nominal thickness of 8 in. Material waste is estimated to be 5%.

Concrete cost = $98/yd³
Concrete pour rate = 8 yd³/hr
Surface finishing compound is purchased in 5-gal containers costing $40 each
Coverage of surfacing compound = 300 ft²/gal

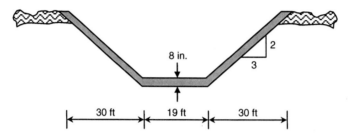

The total material cost ($) is most nearly:

A. 75,000
B. 85,000
C. 95,000
D. 105,000

## 605

Dump trucks with a capacity of 16 yd³ are used to dispose of excavated materials from a site. The distance from the dump-site is 4 mi and the average speed of the dump trucks is 30 mph. The bucket capacity of the power shovel is 3 yd³; power shovel excavation rate is 10 yd³/min (bank measure); the transfer time to the dump trucks (per 3-yd³ load) is 40 sec; and a dump time (per 3-yd³ load) is 30 sec. The excavated material has a swell of 10%.

The factors affecting the job site productivity of the dump truck are: 0.80 for equipment idle time and 0.70 for operator efficiency. If a fleet of dump trucks is used to haul the excavated material, the number of trucks needed is most nearly:

A. 3
B. 4
C. 5
D. 6

## 606

A certain project includes the activities summarized in the table below. The start date of the project is week 0. The project duration (weeks) is most nearly:

A. 32
B. 36
C. 41
D. 44

| Activity | Predecessor Activities | Duration (weeks) |
|---|---|---|
| Start | – | 0 |
| A | Start | 9 |
| B | A | 8 |
| C | B | 2 |
| D | A, C | 6 |
| E | D | 7 |
| F | C, D | 9 |
| G | E, F | 7 |
| Finish | G | 0 |

## 607

A 6-in.-square unreinforced-concrete beam is loaded in third-point loading on a simple span of 5 ft. The test is conducted at 28 days after the concrete beam is cast. If the failure occurs by flexural cracking within the middle third of the beam when load P = 800 lb, the 28-day compressive strength (lb/in.²) of the concrete is most nearly:

A. 6,500
B. 5,500
C. 4,500
D. 3,500

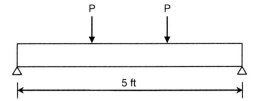

## 608

A 6-in.-diameter × 12-in.-height concrete cylinder fails at a transverse load of 55,000 lb at 28 days in a split cylinder test. The 28-day compression strength (lb/in.²) of the concrete is most nearly:

A. 6,200
B. 5,200
C. 4,200
D. 3,200

## 609

Normal weight concrete (unit weight = 145 lb/ft³) is placed at a rate of 12 ft/hr into 18-ft-high column forms. The blended cement mix contains 30% fly ash but no retarders. Average concrete temperature during curing is 50°F. The maximum lateral design pressure (lb/ft²) for the design of forms is most nearly:

A. 2,600
B. 2,760
C. 2,150
D. 2,300

## 610

A temporary road closure (not exceeding 20 min during the daytime) is established around a 1,250-ft-long work zone (including a buffer space) on a freeway. The distance (ft) from the centerline of the work zone to the location of the ROAD WORK AHEAD sign is most nearly:

A. 4,800
B. 5,800
C. 6,800
D. 7,800

## 611

Whenever a masonry wall is constructed, a limited access zone is to be established. Which of the following is true?

A. The limited access zone shall run to the same length on both sides of the wall.
B. The width of the limited access zone shall be wall height plus 2 ft.
C. The width of the limited access zone shall be wall height plus 4 ft.
D. If provided along the entire length of the wall, the width of the limited access zone is based on the judgment of the contractor.

## 612

According to OSHA, lifelines shall be secured above the point of operation to an anchorage or structural member capable of supporting a dead weight (lb) of at least:
A. 5,000
B. 4,500
C. 4,000
D. 3,200

## 613

Results from a standard Proctor compaction test of six soil samples from a borrow pit are tabulated below. The natural moisture content of the excavated material is 12%. The fill location requires 1.5 million yd³ of soil compacted to a minimum 90% of the maximum Proctor dry density.

| Sample | Net Weight of Soil (lb) | Moisture Content (%) |
|--------|-------------------------|----------------------|
| 1 | 3.24 | 12 |
| 2 | 3.70 | 14 |
| 3 | 3.95 | 16 |
| 4 | 4.21 | 18 |
| 5 | 3.90 | 20 |
| 6 | 3.40 | 22 |

The total volume of borrow soil that must be excavated (yd³) is most nearly:
A. 1.72 million
B. 1.65 million
C. 1.53 million
D. 1.42 million

## 614

According to OSHA, the limiting exposure (dB) to impulse or impact peak noise level is most nearly:

A. 80
B. 100
C. 120
D. 140

## 615

The *Arrhenius function* for concrete maturity is given below:

$$t_{20} = \sum e^{-\frac{E}{R}\left(\frac{1}{T_i} - \frac{1}{T_o}\right)} \Delta t$$

where  $t_{20}$ = time required, when curing at 20°C, to reach an equivalent maturity
  $E$ = activation energy = 41.5 kJ/mol for T ≥ 20°C
  $R$ = universal gas constant = 8.3144 J/(mol-K)
  $T_i$ = average concrete temperature during time interval $\Delta t$, K
  $T_0$ = reference temperature = 293 K

If the average concrete temperature during curing is 28°C, the equivalent age (days) after 3 days of curing is most nearly:

A. 2.2
B. 3.7
C. 4.1
D. 4.7

## 616

A job requires specially engineered fill to be hauled from storage to the job site and then placed and compacted. Given the following information the installed cost ($/yd³) of the fill is most nearly:

A. 15
B. 27
C. 18
D. 22

| Compacted volume required at job site | 15,000 yd³ |
| Distance from storage to job site | 32 mi |
| Average truck speed | 32 mph |
| Truck capacity (net) | 50,000 lb |
| Truck load time | 25 min |
| Truck unload time | 10 min |
| Truck operating cost | $80 per hr |
| Cost of fill at storage facility | $3 per ton |
| Compacted unit weight of fill | 140 lb/ft³ |
| Cost to place & compact fill | $0.50 per yd³ |

## 617

The safety log for a steel fabrication yard is shown below:

| Total number of full-time employees | 400 |
| Average number of hours worked during the year | 1,970 |
| Total number of deaths | 0 |
| Total number of cases with days away from work | 13 cases |
| Total number of days away from work | 120 days |
| Total other recordable cases | 18 cases |

What is most nearly the annual total recordable case rate, as defined by OSHA?

A.  5

B.  8

C.  18

D.  31

## 618

The following table shows areas of cut and fill sections at locations 50 ft apart. Assume shrinkage = 12% and bulking = 25%. Cumulative earthwork (yd³) between stations 0 + 0.00 and 3 + 0.00 is most nearly:

A.  1,340

B.  1,850

C.  2,200

D.  2,500

(*Continued on next page*)

| Station | Area (ft²) | |
| --- | --- | --- |
| | Cut | Fill |
| 0 + 00.00 | 563.2 | 342.2 |
| 0 + 50.00 | 213.5 | 213.6 |
| 1 + 00.00 | 123.5 | 343.3 |
| 1 + 50.00 | 654.6 | 111.0 |
| 2 + 00.00 | 973.1 | 762.4 |
| 2 + 50.00 | 567.3 | 342.9 |
| 3 + 00.00 | 451.6 | 190.4 |

## 619

A contractor has the following options for a project lasting 18 months:

Option A: Monthly rental of excavation equipment at $15,000 per month + operating costs $2,000/month

Option B: Purchase equipment for $200,000

Maintenance costs = $8,000/month
Resale value of equipment after 18 months = $120,000
Nominal interest rate = 10% p.a.

The benefit:cost ratio of option B (purchasing) is most nearly:

A. 0.95
B. 1.25
C. 1.55
D. 1.95

## 620

An excavation plan is outlined in the following figure. Each grid square is 50 ft × 50 ft. The numbers are depth of cut (ft) at indicated locations. The total volume of cut (yd³) is most nearly:

A. 4,500
B. 5,000
C. 5,500
D. 6,000

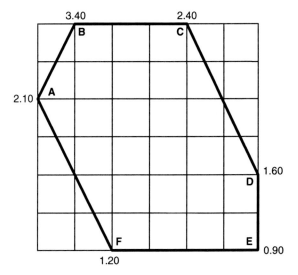

## 621

An existing landfill, in the shape of a truncated pyramid, is rectangular at the base with plan dimensions 4,000 ft × 2,500 ft. The free-standing sides slope 3H:1V. The height of the landfill is 12 ft. The top of the landfill must be covered with cover soil to a depth of 18 in. The amount of cover soil (yd³) required is most nearly:

A. 525,000
B. 650,500
C. 712,200
D. 821,435

## 622

The following table shows ordinates of a mass haul diagram for a highway construction project. The free haul distance is 300 ft. Overhaul cost is $3.70/station-yd.

| Station | 30 + 00 | 42 + 40 | 45 + 10 | 47 + 00 | 48 + 10 | 50 + 30 | 53 + 20 |
|---|---|---|---|---|---|---|---|
| Cumulative Yardage (yd³) | 0 | 800 | 1275 | 1580 | 1275 | 680 | 0 |

*(Continued on next page)*

The total cost of overhaul is most nearly:

A. $35,000
B. $40,000
C. $45,000
D. $50,000

## 623

The total duration of a project was estimated as follows:

522 working days at 6 days per week and 10 hr a day of work.
30 days (10 hr per day) for contingencies such as bad weather, which would be worked on weekends and holidays.

The labor agreement defines the regular work day as an 8-hr-long week-day. Labor cost for premium time is 50% for hours during the week, and 80% for weekend work. What is most nearly the overall labor premium rate for the project?

A. 1.25
B. 1.28
C. 1.31
D. 1.34

## 624

A roller is used to compact soil to a specific density. The roller can achieve this target density for a 6-in. layer of material in four passes. The following specifications are given:

Width of roller drum = 8.0 ft
Forward speed of roller = 3 mph
Overall efficiency of roller = 80%
Soil shrinkage = 15%

The material delivery rate (bank measure) to the site is 950 yd$^3$/hr. The number of rollers required to keep up with this material delivery rate is most nearly:

A. 1
B. 2
C. 3
D. 4

## 625

Specifications for a construction job state that wall forms can be removed once the concrete reaches a compressive strength of 3,400 psi. The maturity curve (Nurse-Saul) for the concrete is shown in the figure below. The cement hydration is assumed to cease below a temperature of 30°F. The average temperature of the concrete during the curing process is 70°F. The number of hours that the contractor has to wait before removing forms is most nearly:

A. 24
B. 36
C. 48
D. 60

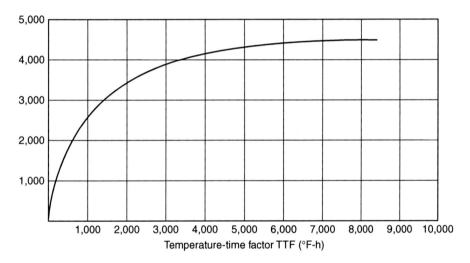

Temperature-time factor TTF (°F-h)

## 626

A 14-ft-high wall is placed at a rate of 4 ft/hr. Concrete is lightweight (135 pcf) made with Type I cement and a retarding admixture. The average concrete temperature is 60°F. The lateral design pressure exerted by the fresh concrete is most nearly:

A. 870
B. 750
C. 900
D. 725

## 627

An activity on an arrow network for a project is shown below. Numbers adjacent to arrows are activity durations (weeks). The total float (weeks) for activity G is:

A. 0

B. 1

C. 2

D. 3

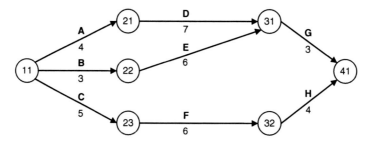

## 628

As part of the PERT analysis of a project, durations are estimated for activities A through G and summarized in the table below. The critical path for the project is ACEF.

| Activity | Duration (weeks) | | |
|---|---|---|---|
| | Optimistic | Most Probable | Pessimistic |
| A | 3 | 4 | 5 |
| B | 4 | 5 | 7 |
| C | 5 | 7 | 8 |
| D | 3 | 4 | 5 |
| E | 5 | 6 | 7 |
| F | 7 | 9 | 10 |
| G | 4 | 5 | 6 |

The probability (%) that the project will be completed in less than 25 weeks is most nearly:

A. 12

B. 18

C. 20

D. 22

## 629

A new earth haul project requires moving 323,000 ft³ (bank measure) of excavated material to a location 12 mi away. A bidding contractor had completed a similar project (size and scope) 2½ years ago. The proposed price for the job on record was $525.00 per dump-truck load. Annual construction inflation factor is 3.2%. The contractor uses a fleet of dump trucks with 26-yd³ capacity (heaped). The excavated material has a swell factor of 25%. The contractor's bid ($) should be most nearly:

A.  262,000
B.  302,000
C.  326,000
D.  353,000

## 630

A crate weighing 900 lb is being lifted by four cables attached to the corners as shown below. The attachment point for the cables (O) is 8 ft directly above the center of gravity of the load. Tension in each cable (lb) is most nearly:

A.  270
B.  225
C.  342
D.  412

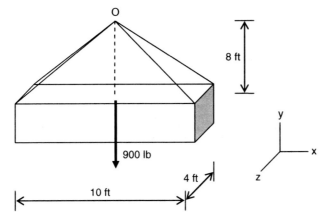

## 631

It is proposed to drive a concrete pile using a pile hammer with energy of 50,000 ft lb. The minimum factor of safety against bearing capacity failure is equal to 6.0. The superstructure load transmitted to the pile is expected to be 40 tons.

The ENR pile driving equation gives the static (ultimate) bearing capacity of a pile based on a pile driving test as:

$$Q_{ult} = \frac{WH}{S+1.0}$$

where $Q_{ult}$ = ultimate capacity (lb)
$W$ = weight of the ram (lb)
$H$ = height the ram falls (in.)
$S$ = pile set (in. per blow)

The required blow count (blows per ft) to achieve a design capacity of 40 tons is most nearly:

A.  8 blows per ft
B.  28 blows per ft
C.  35 blows per ft
D.  50 blows per ft

## 632

A building construction site is represented by the following rectangle ABCD. The original elevation of the water table at the site is 325.8 ft (above sea level). The dashed line shows the limits of proposed construction. The bottom elevation of the mat foundation for the building is 325.65 ft. Steady state pumping from a 9-in.-diameter well at point B is to be used to lower the GWT. The transmissivity of the underlying aquifer is 250 ft²/hr. During construction, the GWT must be lowered to at least 5 ft below the bottom of the foundation and no higher than an elevation of 324.0 ft within the limits of the site. The minimum steady-state pumping rate (gal/min) required is most nearly:

A.  1,000
B.  1,500
C.  2,000
D.  3,000

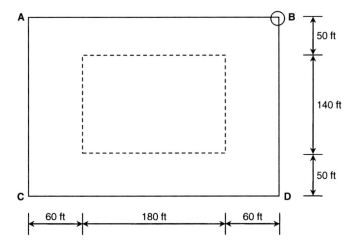

## 633

A retaining wall (height = 24 ft) is shown in the figure below. Lateral loads are resisted by geotextile blankets spaced at 3-ft vertical spacing as shown. The angle of friction between the backfill soil and the reinforcing strips is $\delta = 15°$. The required length (ft) of the reinforcing strip (shown as a solid line) at a depth D = 10 ft below the surface is most nearly:

A. 2
B. 10
C. 8
D. 16

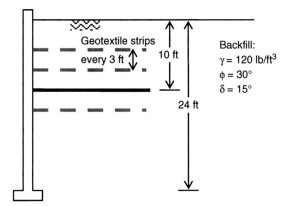

## 634

A contract has 90 days remaining until completion. Early completion will be paid $18,000 bonus per day that the duration is shortened. Activities on the critical path are targeted for "crashing." Critical activities can be accelerated by allocating additional crew units. Additional cost and schedule reduction are shown in the table below. If crews must be added as a complete unit, the net bonus is maximized by the addition of how many crews?

A. 1 additional crew unit

B. 2 additional crew units

C. 3 additional crew units

D. None of the crashing schemes are profitable

| Crew Units Added | Added Crew Cost per Day | Schedule Days Reduced |
|:---:|:---:|:---:|
| 1 | $3,200 | 15 |
| 2 | $6,000 | 21 |
| 3 | $9,000 | 31 |

## 635

A concrete mix has the following components:

| | |
|---|---|
| Cement | 160 lb |
| Wet sand (moisture content = 5%) | 290 lb |
| Wet coarse aggregate (moisture content = 3%) | 420 lb |
| Added water | 56 lb |
| Air | 3% (by volume) |

The following specifications are given:

| | |
|---|---|
| Cement | specific gravity = 3.15 |
| SSD sand (m.c. = 0.7%) | specific gravity = 2.70 |
| SSD coarse aggregate (m.c. = 0.5%) | specific gravity = 2.60 |

The unit weight of the concrete (lb/ft$^3$) is most nearly:

A. 140

B. 142

C. 144

D. 146

## 636

The activity on a node diagram for a project is shown below. All durations are in weeks. All relationships are finish to start unless noted otherwise. The free float (weeks) of activity A is:

A. 0
B. 1
C. 2
D. 3

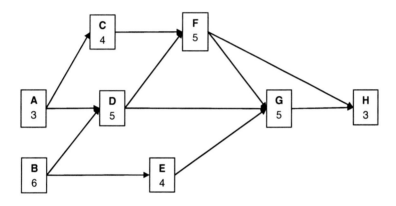

## 637

A construction project is broken down into seven stages (A-G). The sequencing of these stages is represented by a Gantt chart, as shown. The time axis is marked in units of (5-day) weeks. The Gantt chart shows current completion levels (shaded in gray) superimposed on the initially planned timeline. The current status of the project at 4 weeks 2 days is shown. Which of the following statements is true?

A. All activities in stage E must be completed before the initiation of stage G.
B. The only stage that is on or ahead of schedule is G.
C. Of ongoing or completed stages, those with the lowest levels of completion are C and D.
D. Stages A and D are behind schedule.

(*Continued on next page*)

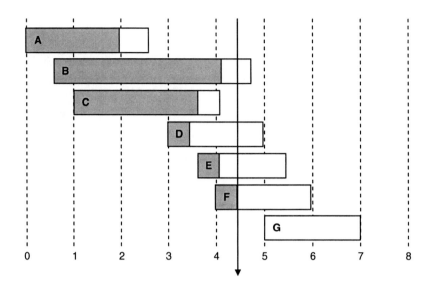

## 638

All costs for a construction project are classified as either direct costs (material, labor, etc.) or indirect costs (leasing and rental costs, utilities, etc.). The estimated variation of these cost categories vs. project duration (weeks) is shown. The optimum project duration (weeks) is most nearly:

A.  10
B.  13
C.  16
D.  19

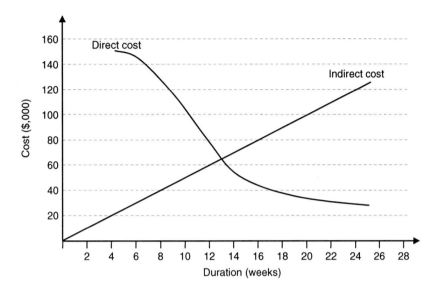

## 639

A 5-story structure steel frame is to be constructed. Story height is 12 ft. The maximum tolerance for out-of-plumbness of the columns (in.) at the fourth floor is most nearly:

A. 0.25
B. 0.65
C. 0.95
D. 1.15

## 640

A crane is used to lift a load of W = 40 tons, as shown. The crane cabin has a ballasted weight of 50 k. Center of gravity is at CG. The allowable soil pressure is 3,800 lb/ft². If the crane is supported by four outriggers as shown, the maximum lateral offset X (ft) is most nearly:

A.  5
B.  8
C.  17
D.  25

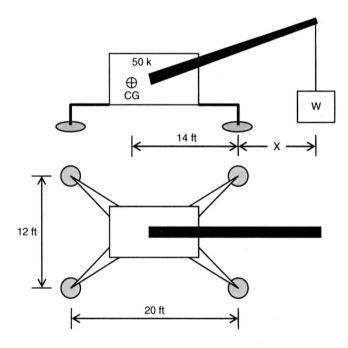

**END OF CONSTRUCTION DEPTH EXAM**

**USE THE ANSWER SHEET ON THE NEXT PAGE**

## Construction Depth Exam: Answer Sheet

| | | | | |
|---|---|---|---|---|
| 601 | Ⓐ | Ⓑ | Ⓒ | Ⓓ |
| 602 | Ⓐ | Ⓑ | Ⓒ | Ⓓ |
| 603 | Ⓐ | Ⓑ | Ⓒ | Ⓓ |
| 604 | Ⓐ | Ⓑ | Ⓒ | Ⓓ |
| 605 | Ⓐ | Ⓑ | Ⓒ | Ⓓ |
| 606 | Ⓐ | Ⓑ | Ⓒ | Ⓓ |
| 607 | Ⓐ | Ⓑ | Ⓒ | Ⓓ |
| 608 | Ⓐ | Ⓑ | Ⓒ | Ⓓ |
| 609 | Ⓐ | Ⓑ | Ⓒ | Ⓓ |
| 610 | Ⓐ | Ⓑ | Ⓒ | Ⓓ |
| 611 | Ⓐ | Ⓑ | Ⓒ | Ⓓ |
| 612 | Ⓐ | Ⓑ | Ⓒ | Ⓓ |
| 613 | Ⓐ | Ⓑ | Ⓒ | Ⓓ |
| 614 | Ⓐ | Ⓑ | Ⓒ | Ⓓ |
| 615 | Ⓐ | Ⓑ | Ⓒ | Ⓓ |
| 616 | Ⓐ | Ⓑ | Ⓒ | Ⓓ |
| 617 | Ⓐ | Ⓑ | Ⓒ | Ⓓ |
| 618 | Ⓐ | Ⓑ | Ⓒ | Ⓓ |
| 619 | Ⓐ | Ⓑ | Ⓒ | Ⓓ |
| 620 | Ⓐ | Ⓑ | Ⓒ | Ⓓ |

| | | | | |
|---|---|---|---|---|
| 621 | Ⓐ | Ⓑ | Ⓒ | Ⓓ |
| 622 | Ⓐ | Ⓑ | Ⓒ | Ⓓ |
| 623 | Ⓐ | Ⓑ | Ⓒ | Ⓓ |
| 624 | Ⓐ | Ⓑ | Ⓒ | Ⓓ |
| 625 | Ⓐ | Ⓑ | Ⓒ | Ⓓ |
| 626 | Ⓐ | Ⓑ | Ⓒ | Ⓓ |
| 627 | Ⓐ | Ⓑ | Ⓒ | Ⓓ |
| 628 | Ⓐ | Ⓑ | Ⓒ | Ⓓ |
| 629 | Ⓐ | Ⓑ | Ⓒ | Ⓓ |
| 630 | Ⓐ | Ⓑ | Ⓒ | Ⓓ |
| 631 | Ⓐ | Ⓑ | Ⓒ | Ⓓ |
| 632 | Ⓐ | Ⓑ | Ⓒ | Ⓓ |
| 633 | Ⓐ | Ⓑ | Ⓒ | Ⓓ |
| 634 | Ⓐ | Ⓑ | Ⓒ | Ⓓ |
| 635 | Ⓐ | Ⓑ | Ⓒ | Ⓓ |
| 636 | Ⓐ | Ⓑ | Ⓒ | Ⓓ |
| 637 | Ⓐ | Ⓑ | Ⓒ | Ⓓ |
| 638 | Ⓐ | Ⓑ | Ⓒ | Ⓓ |
| 639 | Ⓐ | Ⓑ | Ⓒ | Ⓓ |
| 640 | Ⓐ | Ⓑ | Ⓒ | Ⓓ |

# 8

# Breadth Exam No. I
# Solutions

These detailed solutions are for questions 1 to 40, representative of a 4-hour breadth exam according to the syllabus and guidelines for the Principles and Practice (P&P) of Civil Engineering Examination, administered by the National Council of Examiners for Engineering and Surveying (NCEES).

## I

Wetted perimeter: $P = 12 + 2 \times 4.6 = 21.2$ ft

Flow area: $A = 12 \times 4.6 = 55.2$ ft$^2$

Hydraulic radius: $R = \dfrac{A}{P} = \dfrac{55.2}{21.2} = 2.60$ ft

According to Chezy-Manning, average velocity:

$$V = \frac{1.486}{0.014} \times 2.6^{2/3} \times \sqrt{0.004} = 12.69$$

Flow rate: $Q = VA = 12.69 \times 55.2 = 700$ cfs    **B**

(*Continued on next page*)

### Alternate solution

Using tables for flow parameter $K$ (see Table 303.3 in I. Goswami, *Civil Engineering PE All-in-One Exam Guide: Breadth and Depth*, McGraw-Hill, 2009)

For depth ratio $\dfrac{d}{b} = \dfrac{4 \cdot 6}{12} = 0.383 \rightarrow K = 0.1385$, Table 303.3 gives $K = 0.1385$

Flow rate: $Q = \dfrac{1.486}{0.014} \times 0.1385 \times 12^{8/3} \times \sqrt{0.004} = 701.8$ cfs (*All-in-One* Equation 303.36)

## 2

Since flow occurs by gravity alone, application of Bernoulli's equation between the two reservoir surfaces will yield:

$$z_1 + \frac{V_1^2}{2g} + \frac{p_1}{\gamma} - h_f = z_2 + \frac{V_2^2}{2g} + \frac{p_2}{\gamma}$$

Since the two reservoir surfaces have negligible velocity and both are at atmospheric pressure, $V_1 = V_2 \simeq 0$ and $p_1 = p_2 = p_{atm}$

$$h_f = z_1 - z_2 = 70 \text{ ft}$$

Equivalent length of pipe: $L_{eq} = 2{,}500 + 55 = 2{,}555 \text{ ft}$

$$h_f = 70 = f\frac{L}{D}\frac{V^2}{2g} = 0.02 \times \frac{2{,}555}{2} \times \frac{V^2}{64.4} \Rightarrow V = 13.3 \text{ fps}$$

Flow rate: $Q = V \times A = 13.3 \times \dfrac{\pi}{4} \times 2^2 = 41.73 \text{ cfs}$     **A**

## 3

The flow rate at the end of the second hour is calculated as the superposition of the effect of the first hour (with a 2-hr lag because the first hour of excess precipitation STARTS at $t = 0$) and the effect of the second hour (with a 1-hr lag). Each of these ordinates is scaled by the appropriate depth of excess precipitation (1.7 in. and 0.8 in., respectively).

$$Q_3 = 95 \times 1.7 + 30 \times 0.8 = 185.5 \text{ cfs}$$     **D**

## 4

The longitudinal slope is calculated as the invert elevation difference divided by the length of the pipe:

$$S = \frac{\Delta z}{L} = \frac{275.64 - 270.96}{800} = 0.00585$$

Flow rate in circular pipe flowing full (*All-in-One* Equation 303.34)

$$Q_f = \frac{0.312 \times 1.486}{n} D^{8/3} S^{1/2} = \frac{0.464}{0.013} \times 2.5^{8/3} \times \sqrt{0.00585} = 31.43 \text{ cfs}$$

Using Table 303.2 from *All-in-One*

Flow ratio: $\dfrac{Q}{Q_f} = \dfrac{20}{31 \cdot 43} = 0.64 \rightarrow \dfrac{d}{D} = 0.58 \rightarrow d = 0.58 \times 30 = 17.4$ in.  **C**

## 5

For horizontal backfill, and ignoring friction between the back of the wall and backfill (i.e., using Rankine's theory): $\phi = 34 \rightarrow K_a = 0.283$

Height of active zone = 257 − 240 = 17 ft

Resultant earth pressure force: $R_a = \dfrac{1}{2} \times 0.283 \times 120 \times 17^2 = 4{,}900 \text{ lb/ft}$

Total gravity force (weight of concrete wall components and soil above the heel) is $W = (11 \times 3 + 14 \times 1) \times 150 + 6 \times 14 \times 120 = 17{,}130 \text{ lb/ft}$

Available friction force: $F_f = W \tan 20 = 6{,}234.8 \text{ lb/ft}$

Factor of safety for sliding instability: $FS = \dfrac{6235}{4900} = 1.27$  **A**

## 6

Soil sample unit weight: $\gamma = \dfrac{3.64}{0.031} = 117.4 \text{ lb/ft}^3$

The volume of water added until all voids are filled = 5.6 oz. This must be the volume of air in the original sample.

$V_{air} = 5.6 \text{ oz.} = 0.00585 \text{ ft}^3$ (128 fl. oz. = 1 US gal = 0.13368 ft³)

(*Continued on next page*)

Therefore, the remaining volume is occupied by soil solids and water:

$$V_s + V_w = 0.031 - 0.00585 = 0.02515 \text{ ft}^3 \qquad \text{(a)}$$

Since only these two constituents contribute to the soil weight, we write the expression for sample weight:

$$2.65 \times 62.4 \times V_s + 62.4 \times V_w = 3.64 \qquad \text{(b)}$$

If we solve equations (a) and (b), we get:

$$V_s = 0.020 \text{ ft}^3$$

Weight of solids: $W_s = 2.65 \times 62.4 \times 0.02 = 3.326 \text{ lb}$

Dry unit weight of the soil: $\gamma_d = \dfrac{3.326}{0.031} = 107.3 \text{ lb/ft}^3$      **C**

## 7

The following are Terzaghi's values of the bearing capacity factors or friction angle $\phi = 10^0$:

$$N_c = 9.6; \; N_q = 2.7; \; N_\gamma = 0.56$$

For a soil with a small angle of internal friction, cohesion is approximately half the unconfined compression strength = 1,200 lb/ft²

For a square footing, Terzaghi's bearing capacity equation yields:

$$q_{ult} = 1.3cN_c + \gamma DN_q + 0.4\gamma BN_\gamma$$

$$q_{ult} = 1.3 \times 1200 \times 9.6 + 122 \times 3 \times 2.7 + 0.4 \times 122 \times B \times 0.56 = 15,964 + 27B$$

$$q_{all} = \frac{a_{ult}}{FS} = \frac{15,964 + 27B}{3} = 5,321 + 9B$$

Allowable bearing load (lb) must be greater than the applied column load:

$$(5,321 + 9B)B^2 \geq 80,000 \Rightarrow B \geq 3.87 \text{ ft} \quad \text{(by trial and error)} \qquad \textbf{C}$$

## 8

Volatile suspended solids concentration: $VSS = 0.7 \times 1,100 = 770$ mg/L

Volatile suspended solids mass transfer rate (lb/day): $4 \times 770 \times 8.3454 = 25,704$ lb/day

*Note:* Conversion factor 8.3454 lb-L/mg-MG

Tank volume: $V = 25,000 \text{ gal} = \dfrac{25,000}{7.48} = 3,342 \text{ ft}^3$

Organic load $= \dfrac{25,704}{3,342} = 7.7 \text{ lb} \cdot \text{VSS/ft}^3 \cdot \text{day}$ 　　　　　　　　　D

## 9

Maximum moment: $M = \dfrac{wL^2}{8} = \dfrac{2 \cdot 75 \times 25^2}{8} = 214.8 \text{ k} \cdot \text{ft} = 2,578 \text{ k} \cdot \text{in.}$

Required section modulus: $S_{reqd} \geq \dfrac{M}{\sigma_{all}} = \dfrac{2,578}{32} = 80.6 \text{ in.}^3 \rightarrow W12 \times 65$ 　　　D

## 10

The load on the beam has three zones. In the first, from the free end on the left to the left support, the load is zero. In the second zone, on the right of the left support, the load function is constant (uniformly distributed). In the third zone, the load is again zero. According to the geometric principles, the *bending moment function is two orders higher than the load function.* Therefore, the bending moment diagram should have a LINEAR segment followed by a QUADRATIC segment, followed by another LINEAR segment. The only one to have this pattern is 　　　　　　　　D

## 11

The I-shaped section is divided into three rectangles—bottom flange, web, and top flange. Elastic neutral axis (centroid) location (using bottom edge of section as datum) is given by:

$$\bar{y} = \frac{\sum y_i A_i}{\sum A_i} = \frac{0.5 \times 8 + 6 \times 10 + 11.25 \times 3}{8 + 10 + 3} = 4.655$$

*(Continued on next page)*

Using the Parallel Axis Theorem, the moment of inertia $I_{xc}$ is given by:

$$I_{xc} = \frac{1}{12} \times 6 \times 0.5^3 + 3 \times (11.25 - 4.655)^2 + \frac{1}{12} \times 1 \times 10^3 + 10 \times (6 - 4.655)^2 + \frac{1}{12} \times 8$$

$$\times 1^3 + 8 \times (0.5 - 4.655)^2 = 370.74 \text{ in.}^4$$

The farthest fiber from the neutral axis is at a distance $c = 11.5 - 4.655 = 6.845$ in.

Elastic section modulus: $S_x = \dfrac{I_{xc}}{c} = \dfrac{370.74}{6.845} = 54.2 \text{ in}^3$     **A**

## 12

Half the curve length of the distance from the PVC to the PV is:

$$\frac{L}{2} = 123.645 - 117.5 = 6.145 \text{ sta}$$

Then the elevation of the PVT can be calculated using the slope of the forward tangent from the PVI is:

$$y_{PVT} = 325.64 - 3 \times 6.145 = 307.21 \text{ ft}$$     **A**

## 13

To calculate the mean headway, one needs only the flow rate. The headway is calculated as:

$$h_a = \frac{3{,}600 \text{ sec/hr}}{1{,}450 \text{ veh/hr}} = 2.48 \text{ s/veh}$$     **D**

## 14

The superelevation factor is calculated as $e + f = 0.04 + 0.08 = 0.12$

With design speed $V$ (mph) and curve radius $R$ (ft), the following is the condition for lateral stability:

$$e + f \geq \frac{V^2}{15R} \Rightarrow R \geq \frac{V^2}{15(e+f)} = \frac{40^2}{15 \times 0.12} = 889 \text{ ft}$$     **A**

This criterion states that the MINIMUM curve radius is 889 ft. Therefore, one must pick the smallest value above 889 ft and not the nearest.

## 15

For a design speed of 45 mph on a 5% downgrade, the stopping sight distance is:

$$SSD = 1.47 \times 2.5 \times 45 + \frac{1.075 \times 45^2}{11.2 - 0.05 \times 32.2} = 392.4 \text{ ft}$$

In the equation above, we use the default values from the AASHTO Green Book for reaction time ($t_R = 2.5$ sec) and deceleration rate ($a = 11.2$ ft/s²).

The Green Book rounds it to 395 ft                                                                A

## 16

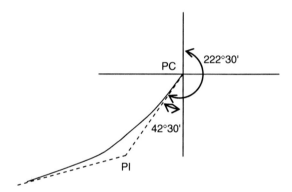

Back tangent azimuth 42.5 + 180 = 222.5 (measured clockwise from North meridian)

$$R = 1,030 \text{ ft}; L = 646.35 \text{ ft}; I = \frac{180L}{\pi R} = 35.95°$$

Tangent length, $T = R \tan \frac{I}{2} = 1030 \times \tan \frac{35.95}{2} = 334.22 \text{ ft}$

Change in northing between PC and PI is calculated as:
$\Delta N = T \cos Az = 334.22 \times \cos 222.5 = -246.41 \text{ ft}$

Change in easting between PC and PI is calculated as:
$\Delta E = T \sin Az = 334.22 \times \sin 222.5 = -225.80 \text{ ft}$

Therefore, coordinates (north and east positive) of the PI are:

Northing = 4,123.64 − 2,46.41 = 3,877.23

Easting = −1,064.32 − 225.80 = −1,290.12

(*Continued on next page*)

This can be expressed as (3,877.23 N, 1,290.12 W).                    A

*Note:* Since all answers have unique answer choices for both northings and eastings, calculating just one of them is adequate to identify the correct answer.

# 17

According to the chart reproduced here, flow rate is maximum (greater than $Q_f$) at approximately $d/D = 0.90$ for Manning's $n$ constant with depth and approximately $d/D = 0.95$ for Manning's $n$ variable with depth. We are considering both possibilities because the question does not specify whether to assume $n$ = constant or variable. This means $d = 38$ in. in the first case or $d = 40$ in. in the second case. Only one answer ($d = 40$ in.) fits.                    C

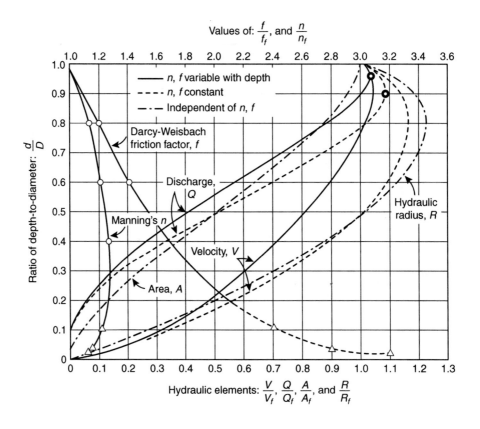

## 18

From the 5-day BOD test, the $BOD_5$ is calculated as:

$$BOD_5 = \frac{6.3 - 2.7}{15/300} = 72 \text{ mg/L}$$

The ultimate BOD, according to the exponential model, is:

$$BOD_u = \frac{BOD_5}{1 - 10^{-kt}} = \frac{72}{1 - 10^{-0.1 \times 5}} = 105.3 \text{ mg/L} \qquad \textbf{B}$$

## 19

Opening size for no. 200 sieve = 0.075 mm. From the particle size distribution curve, the % passing corresponding to this size is $F_{200} = 12\%$. Since this is less than 50%, soil is predominantly coarse grained. First letter is S or G. This eliminates choices C and D. The other two choices are both "sand" (first letter S); therefore, given these answer choices, we do not need to distinguish between G or S.

Percentage passing no. 4 sieve (sieve size 4.75 mm), $F_4 = 98\%$. % retained $R_4 = 2\%$. This is less than half of the coarse fraction = 100 − 12 = 88%. Soil is predominantly sand (not gravel). First letter is S.

The three points marked on the gradation curve correspond to 10%, 30%, and 60% passing are:

$D_{10} = 0.065$ mm, $D_{30} = 0.18$ mm, $D_{60} = 0.5$ mm

$$C_u = \frac{D_{60}}{D_{10}} = \frac{0.5}{0.065} = 7.7 \qquad \text{and} \qquad C_z = \frac{D_{30}^2}{D_{10}D_{60}} = \frac{0.18^2}{0.065 \times 0.5} = 1.0$$

Criteria for well-graded sand are approximately met. The only available choice that fits is SW  $\qquad\qquad$ **A**

## 20

From the particle size distribution curve: $F_{200} = 12\%$

Since this is significant, AASHTO's equivalent to the Casagrande Plasticity Chart is used with LL = 34 and PI = 34 − 19 = 15

(*Continued on next page*)

Table 203.5    AASHTO Soil Classification Criteria

| Sieve Analysis | A-1 | | A-3 | A-2 | | | | A-4 | A-5 | A-6 | A-7 | A-8 |
| | A-1-a | A-1-b | | A-2-4 | A-2-5 | A-2-6 | A-2-7 | | | | | |
|---|---|---|---|---|---|---|---|---|---|---|---|---|
| | Granular Materials (35% or less passing no. 200 sieve) | | | | | | | Silt-Clay Materials (more than 35% passing no. 200 sieve) | | | | |
| % passing | | | | | | | | | | | | |
| No. 10 | ≤50 | | | | | | | | | | | |
| No. 40 | ≤30 | ≤50 | >50 | | | | | | | | | |
| No. 200 | ≤15 | ≤25 | ≤10 | ≤35 | ≤35 | ≤35 | ≤35 | >35 | >35 | >35 | >35 | |
| LL | | | | ≤40 | >40 | ≤40 | >40 | ≤40 | >40 | ≤40 | >40 | |
| PI | ≤6 | | NP | ≤10 | ≤10 | >10 | >10 | ≤10 | ≤10 | >10 | >10 | |
| General description | Stone, gravel, sand | | Fine sand | Silty or clayey gravel and sand | | | | Silty soils | | Clayey soils | | Peat, highly organic soils |
| Quality as subgrade material | Good to excellent | | | | | | | Fair to poor | | | | Very poor |

Soil classification is A2-6 ($F_{200}$ < 35%)

Group Index: $GI = 0.01(F_{200} - 15)(PI - 10) = -0.15$. This should be reported as non-negative. Therefore, $GI = 0$. The calculation of the group index is *shown but not necessary* for this problem. Also, note that the single-term expression above for the group index is recommended for groups A2-6 and A2-7. For other groups, the expression for the group index has an additional term.

Soil is classified as A2-6 (0)                                                              **B**

# 21

A 3-hinged arch is statically determinate (four external reactions $A_x$, $A_y$, $C_x$, and $C_y$) and four equations of static equilibrium (three equations of equilibrium for the overall structure and one hinge moment equation).

The zero moment at internal hinge (considering only the left substructure) is:

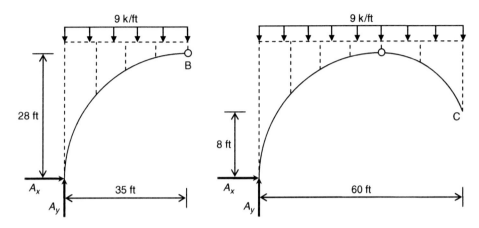

$$\sum M_{B,L} = 0 \Rightarrow 9 \times 35 \times 17 \cdot 5 + 28A_x - 35A_y = 0$$

For the entire structure, the moment equilibrium equation (sum of moment about C) is:

$$\sum M_C = 0 \Rightarrow 9 \times 60 \times 30 + 8A_x - 60A_y = 0$$

Solving these two equations, we get: $A_x = 168.75$ k (to the right)　　　　　　**D**

## 22

For each sack of cement (weight = 94 lb)

Sand (SSD): Weight = 1.8 × 94 = 169.2 lb (this represents sand with 0.5% moisture)

When wet sand (m.c. = 6%) is used, it contains free water = 5.5/100.5 × 169.2 = 9.26 lb

Coarse aggregate (SSD): W = 2.6 × 94 = 244.4 lb (this represents aggregate with 0.7% moisture)

When wet aggregate (m.c. = 4%) is used, it contains free water = 3.3/100.7 × 244.4 = 8.01 lb

Total extra water = 9.26 + 8.01 = 17.27 lb = 2.07 gal (8.3454 lb/gal $H_2O$)

Total water = 5.8 + 2.07 = 7.87 gal/sack　　　　　　**C**

## 23

Since the outriggers are simply resting on the soil (not anchored), the limiting condition is when the far side (left) outrigger legs have zero reaction. For this condition, the inside legs carry the total load of 40 tons. (Incidentally, the offset X for which this occurs is 15 ft, from moment equilibrium).

Therefore, maximum load per leg = 20 tons = 40,000 lb. Based on the allowable soil pressure of 2,800 psf, the outrigger pads must have a minimum area = 40/2.8 = 14.3 sq. ft (each). In this case, choose a bigger pad.    C

## 24

Activity on node version of the network is shown below. Each task is labeled with its duration in the subscript.

Solution progresses as follows (using the notation ES: early start; EF: early finish).

Activity A: $ES_A = 0$; $EF_A = 5$

Activity D: has only one predecessor (A). $ES_D = 5$; $EF_D = 9$

Activity G: FF lag = 5 with D. $EF_G = 14$

Activity J: has multiple predecessors (F, G, I), whose EF times are (not shown) 13, 14, and 13, respectively. Therefore, $ES_J = 14$; $EF_J = 16$

The critical path is ADGJ (length 16 months). Answer is    B

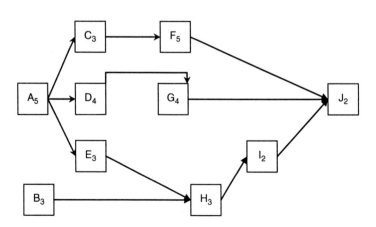

## 25

The critical sight distance is on the inside lane, where the edge offset is 12 ft. The radius of the centerline of the inside lane is calculated by subtracting a half-lane width from the centerline radius.

$$R = 915 - 6 = 909 \text{ ft}$$

For this curve, the middle ordinate distance is calculated by adding a half-lane width to the edge offset.

$$M = 12 + 6 = 18 \text{ ft}$$

The stopping sight distance is the length of curve corresponding to these values of R and M.

$$SSD = \frac{R}{28.65}\left[1 - \cos^{-1}\left(\frac{M}{R}\right)\right] = \frac{909}{28.65} \times \left[1 - \cos^{-1}\left(\frac{18}{909}\right)\right] = 362 \text{ ft}$$

The corresponding design speed (AASHTO Green Book) is 45.2 mph.
Use 45 mph.                                                                                    A

## 26

In this solution, the elevation datum is taken at the impermeable layer. The total head at the upstream ground surface = 40 ft (gage pressure head) + 50 ft (elevation head) = 90 ft. Total head at downstream ground surface (under 5 ft of water) = 5 + 50 = 55 ft. Head loss through seepage = 90 − 55 = 35 ft. The total head at intermediate location X can be expressed as the weighted average (10.5 pressure drops out of 13) shown below:

$$TH_X = \frac{10 \cdot 5}{13} \times 55 + \frac{2 \cdot 5}{13} \times 90 = 61.73 \text{ ft}$$

Alternatively, one can also calculate the total head at X as the total head at upstream ground minus the head loss for 10.5 pressure drops.

$$TH_X = 90 - 35 \times \frac{10 \cdot 5}{13} = 61.73 \text{ ft}$$

(*Continued on next page*)

At point X, the elevation head is: $EH_X = 12$ ft

Therefore, the pressure head is: $PH_X = TH_X - EH_X = 61.73 - 12 = 49.73$ ft

This is equivalent to a pressure: $p_X = 49.73 \times 62.4 = 3,103$ psf $= 21.55$ psi          **D**

## 27

Designating cable forces in AB, BC, and CD as $T_1$, $T_2$, and $T_3$, respectively, the equations of equilibrium at node C are:

$$\sum F_x = 0 \Rightarrow \frac{3}{\sqrt{10}} T_2 = \frac{1}{\sqrt{5}} T_3 \Rightarrow 0.9487 T_2 = 0.4472 T_3 \Rightarrow T_3 = 2.1214 T_2$$

$$\sum F_y = 0 \Rightarrow \frac{1}{\sqrt{10}} T_2 + \frac{2}{\sqrt{5}} T_3 = 3 \text{ tons} \Rightarrow 0.3162 T_2 + 0.8944 \times 2.1214 T_2 = 3 \text{ tons} \Rightarrow T_2$$

$$= 1.355 \text{ tons}$$

Writing the equilibrium equations at B:

$$\sum F_x = 0 \Rightarrow \frac{1}{\sqrt{2}} T_1 = \frac{3}{\sqrt{10}} T_2 \Rightarrow 0.7071 T_1 = 0.9487 T_2 \Rightarrow T_1 = 1.818 \text{ tons}$$

$$\sum F_y = 0 \Rightarrow \frac{1}{\sqrt{2}} T_1 - \frac{1}{\sqrt{10}} T_2 - F = 0 \Rightarrow F = 0.7071 T_1 - 0.3162 T_2 = 0.86 \text{ ton}$$

$$= 1,714 \text{ lb}$$          **B**

## 28

Using Terzaghi's bearing capacity factors (assuming water table too deep to matter, which is true if footing size is less than 10 ft):

$$\phi = 32° \rightarrow N_c = 44.04; \ N_q = 28.52; \ N_\gamma = 26.87$$

Ultimate bearing capacity (square footing) is given by:

$$q_{ult} = 1.3 c N_c + \gamma D N_q + 0.4 \gamma B N_\gamma$$

$$q_{ult} = 0 + 125 \times 3 \times 28.52 + 0.4 \times 125 \times 26.87 B = 10,695 + 1,343.5 B$$

Since minimum factor of safety for bearing capacity = 3.0, the allowable soil pressure is:

$$q_{all} = \frac{a_{ult}}{FS} = 3,565 + 447.8B$$

In terms of pressures (lb/ft²), pressure due to column load + overburden pressure must be less than allowable pressure.

$$(3,565 + 447.8B) \geq \frac{150,000}{B^2} + 125 \times 3 \Rightarrow (3,190 + 447.8B) \geq \frac{150,000}{B^2}$$

$$\Rightarrow B = 5.21 \text{ ft} \qquad\qquad\qquad \text{C}$$

(The equation above is cubic in nature. The most time-efficient method to solve is by plugging in available answer choices.)

## 29

Based on the given data, the cross section at station 10 + 0.00 may be plotted as shown.

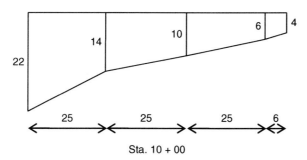

Sta. 10 + 00

Area at station 10 + 00:

$$A = 0.5 \times [(22+14) \times 25 + (14+10) \times 25 + (10+6) \times 25 + (6+4) \times 6] = 980 \text{ ft}^2$$

Similarly, area at station 11 + 00:

$$A = 0.5 \times [(20+15) \times 22 + (15+9) \times 25 + (9+7) \times 25 + (7+5) \times 8] = 933 \text{ ft}^2$$

Volume between stations 10 + 00 and 11 + 00:

$$V = 0.5 \times (980 + 933) \times 100 = 95,650 \text{ ft}^3 = 3,543 \text{ yd}^3 \qquad\qquad \text{B}$$

## 30

OSHA standards require ladders when the trench depth exceeds 4 ft. In that case, no point within the trench must be more than 25 ft distant from a ladder. Placing ladders at $x = 25$ and $x = 70$ ft ensures that all points within the trench are within 25 ft of a ladder. (The most remote point within this gap is at midway between the two ladders at $x = 47.5$, which is 22.5 ft from an egress point.)

Using two ladders satisfies the OSHA requirement for excavations.   **C**

## 31

Cycle time = 45 sec + 4 min + 30 sec = 5 min 15 sec = 5.25 min

50,000 ft$^3$ of soil (bank measure) is equivalent to $50,000 \times 1.2 = 60,000$ ft$^3$ = 2,222 yd$^3$ loose soil

Production (loose soil) in an 8-hr day is calculated as (accounting for 85% efficiency):

$$\frac{480 \text{ min}}{\text{day}} \times \frac{1 \text{ cycle}}{5.25 \text{ min}} \times \frac{2.8 \text{ yd}^3}{\text{cycle}} \times 0.85 = 217.6 \text{ yd}^3/\text{day}$$

Number of days needed = 2,222.2/217.6 = 10.2 days. Choose 11 days to complete.   **B**

## 32

A section through members CD, DI, and HI is shown. To solve for $F_{DI}$, the most efficient method may be to take moments about F (where the other two forces intersect). Note that this means that calculating the support reaction at F is also unnecessary.

Height DH $= \dfrac{2}{3} \times 6 = 4$ ft

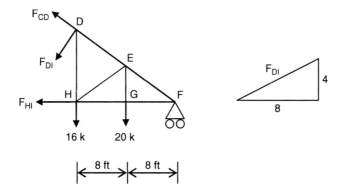

Taking moments about F for the section shown:

$$F_{DI} \cdot \frac{1}{\sqrt{5}} \times 16 + F_{DI} \cdot \frac{2}{\sqrt{5}} \times 4 + 16 \times 16 + 20 \times 8 = 0$$

This leads to $F_{DI} = -38.76$ k (Compression)                                  C

## 33

Point on curve (station 12 + 00) can have maximum elevation = 470.00 − 14.5 = 455.5 ft

Since the PVI is completely specified (station as well as elevation), determine the coordinates of this point on the curve with respect to PVI:

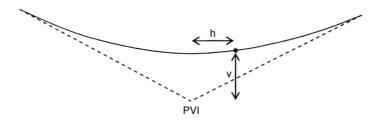

Horizontal offset   $h = 12.0 − 10.563 = 1.437$ stations

Vertical offset      $v = 455.5 − 432.65 = 22.85$ ft

*(Continued on next page)*

The following equation is useful when the PVI is used as the reference point:

$$\frac{L+2h}{L-2h} = \sqrt{\frac{v-G_1 h}{v-G_2 h}}$$

$$\frac{L+2 \times 1.437}{L-2 \times 1.437} = \sqrt{\frac{22.85-(-4)\times 1.437}{22.85-(+6)\times 1.437}} = \sqrt{\frac{28.598}{14.228}} = 1.41774$$

Solving: L = 16.634 sta. = 1,663.4 ft                                        A

*Note:* This is the MINIMUM length of curve. A shorter curve will not provide the necessary vertical clearance. So, if the answer choices were (A) 1,660, (B) 1,700, (C) 1,740, (D) 1,780, the answer would be 1,700 ft in spite of choice (A) 1,660 ft being closest to the solved value of 1,663.4 ft.

## 34

Converting all costs and benefits to present worth.

Initial cost = $75,000 (as is)

Years 1–15 maintenance cost: benefit (savings) = $7,000 annually × (P/A, 15 years, 8%) = 7,000 × 8.5595 = $59,916.50

Years 16–25 maintenance cost: benefit (savings) = $10,000 annually × (P/A, 10 years, 8%) × (P/F, 15 years, 8%) = 10,000 × 6.7101 × 0.3152 = $21,150.24

*Note:* This has been calculated as a shifted annuity. The first factor (P/A, 10 years, 8%) converts the series from years 16–25 into a single amount at the origin of this series (i.e., one period back from the start, at t = 15). The next factor (P/F, 15 years, 8%) converts this sum from year 15 to year 0. This term could also be calculated as:

$10,000 × [(P/A, 25 years, 8%) − (P/A, 15 years, 8%)] = 10,000 × (10.6748 − 8.5595) = $21,153

Salvage value (additional $15,000): future value (considered as negative cost) = $15,000 × (P/F, 25 years, 8%) = 15,000 × 0.1460 = $2,190

Benefit: cost ratio is given by,

$$\frac{B}{C} = \frac{59,916 \cdot 50 + 21,150.24}{75,000 - 2,190} = 1.11$$                    D

## 35

Crew labor rate (per labour hr):

$$\frac{1 \times 30 + 2 \times 18 + 1 \times 12}{4} = \$19.50/\text{L.H.}$$

Area: $A = 2 \times (35+25) \times 14 + 35 \times 25 - 85 = 2{,}470 \text{ ft}^2$

Labor cost:

$$2{,}470 \text{ ft}^2 \div \frac{150 \text{ ft}^2}{LH} \times \frac{\$19.50}{LH} + 2{,}470 \text{ ft}^2 \div \frac{50 \text{ ft}^2}{LH} \times \frac{\$19.50}{LH} = \$ 1{,}284.40 \qquad \text{B}$$

## 36

For strip footing on cohesionless sand ($c = 0$), ultimate bearing capacity is given by:

$$q_{\text{ult}} = \gamma D N_q + \frac{1}{2} \gamma B N_\gamma$$

For $\phi = 30°$, Terzaghi's bearing capacity factors are: $N_q = 22.5$, $N_\gamma = 19.7$

Case A: Footing on ground surface ($D = 0$) is given by:

$$q_{\text{ult}} = \frac{1}{2} \gamma B N_\gamma = \frac{1}{2} \times 120 \times 4 \times 19.7 = 4{,}728 \text{ psf}$$

Case B: Footing depth 30 in. ($D = 2.5$ ft) is given by:

$$q_{\text{ult}} = \gamma D N_q + \frac{1}{2} \gamma B N_\gamma = 120 \times 2.5 \times 22.5 + \frac{1}{2} \times 120 \times 4 \times 19.7 = 11{,}478 \text{ psf}$$

Capacity increase factor $= \dfrac{11{,}478}{4{,}728} = 2.43$ (143% increase) \qquad C

## 37

For site A, degree of consolidation $= 8/10 = 0.8$; time factor $T_v = 0.567$.

Since the clay layer has well-draining sand layers above and below it, the drainage thickness is half the layer thickness, i.e., $H_d = 5$ ft

*(Continued on next page)*

Time for 80% consolidation = 5 years

$$t = \frac{T_v H_d^2}{C_v} \Rightarrow C_v = \frac{T_v H_d^2}{t} = \frac{0.567 \times 5^2}{5} = 2.835 \frac{ft^2}{year}$$

For site B, with single drainage, $H_d = 10$ ft

$$t = \frac{T_v H_d^2}{C_v} \Rightarrow T_v = \frac{C_v t}{H_d^2} = \frac{2.835 \times 5}{10^2} = 0.142$$

Since the clay layer is identical and the pressure conditions are identical, the ultimate consolidation settlement is the same in both cases = 10 in.

Corresponding to $T_v = 0.142$; degree of consolidation $U = 44\%$; expected settlement = $0.44 \times 10 = 4.4$ in.                                                             **B**

## 38

Waste produced by the original population of town A = $20,000 \times 5 = 100,000$ lb/day

At a compacted density of 40 lb/ft$^3$, the volume taken up in the landfill = $\frac{100,000}{40}$ = 2,500 ft$^3$/day = 912,500 ft$^3$/year = 33,796 yd$^3$/year

Remaining capacity = $\frac{1 \times 10^6}{33,796} = 29.6$ years

Waste produced by expanded population of town A = $25,000 \times 5 = 125,000$ lb/day

At a compacted density of 40 lb/ft$^3$, the volume taken up in the landfill = $125,000 \div 40 = 3125$ ft$^3$/day = 1,140,625 ft$^3$/year = 42,245 yd$^3$/year

Remaining capacity = $\frac{1 \times 10^6}{42245} = 23.7$ years

Reduction in service life = $29.6 - 23.7 = 5.9$ years                                     **A**

## 39

The correct pattern is the one that has steel in the areas developing tensile stress due to the wall being loaded laterally (to the left). Answer is **B**

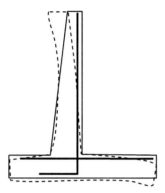

## 40

Hypotenuse $AB = 15$ ft

Resultant of the distributed load $= 60$ k

For the entire frame, taking moments about A:

$$\sum M_A = 0 \Rightarrow 60 \times 7.5 + 25 \times 9 - D_y \times 26 = 0 \Rightarrow D_y = 25.96 \text{ k}$$

Now, make a cut at B and consider the equilibrium of the substructure on the right:

$$\sum M_B = 0 \Rightarrow M_B - 25.96 \times 14 = 0 \Rightarrow M_B = 363.44 \text{ k} \cdot \text{ft} \qquad \text{D}$$

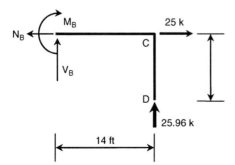

# Answer Key for Breadth Exam No. 1

| | | | | | | | | | |
|---|---|---|---|---|---|---|---|---|---|
| 1 | B | 11 | A | 21 | D | 31 | B |
| 2 | A | 12 | A | 22 | C | 32 | C |
| 3 | D | 13 | D | 23 | C | 33 | A |
| 4 | C | 14 | A | 24 | B | 34 | D |
| 5 | A | 15 | A | 25 | A | 35 | B |
| 6 | C | 16 | A | 26 | D | 36 | C |
| 7 | C | 17 | C | 27 | B | 37 | B |
| 8 | D | 18 | B | 28 | C | 38 | A |
| 9 | D | 19 | A | 29 | B | 39 | B |
| 10 | D | 20 | B | 30 | C | 40 | D |

# 9

# Breadth Exam No. 2 Solutions

These detailed solutions are for questions 101 to 140, representative of a 4-hour breadth exam according to the syllabus and guidelines for the Principles and Practice (P&P) of Civil Engineering Examination, administered by the National Council of Examiners for Engineering and Surveying (NCEES).

## 101

For the entire structure, taking moments about $E$:

$$\sum M_E = 0 \Rightarrow A_y \times 4h + 6 \times \frac{h}{2} + 4 \times h - 24 \times 3h - 40 \times 2h - 12 \times h = 0$$

$$A_y = \frac{157}{4} = 39.25 \text{ k}$$

Reactions are: $A_x = -10$ k, $A_y = 39.25$ k. The forces acting on joint A are shown next.

*(Continued on next page)*

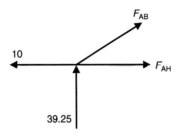

Method of joints at A yields:

$$\sum F_y = 0 \Rightarrow 39.25 + \frac{1}{\sqrt{5}} F_{AB} = 0 \Rightarrow F_{AB} = -87.77 \text{ k}$$

$$\sum F_x = 0 \Rightarrow -10 + \frac{2}{\sqrt{5}} F_{AB} + F_{AH} = 0 \Rightarrow F_{AH} = +88.5 \text{ k}$$

$F_{AH} = 88.5 \text{ k (tension)}$                                                                    B

*Note*: For a node such as A, where three forces intersect, the polygon of forces is a triangle. Therefore, the force triangle will be similar to the geometric triangle. The free body diagram at node A is shown below.

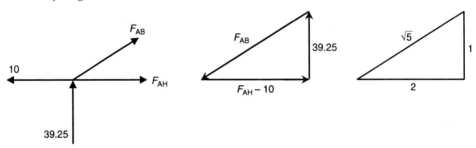

This leads to the following similarity relations:

$$\frac{F_{AH} - 10}{2} = \frac{39.25}{1} = \frac{F_{AB}}{\sqrt{5}} \quad \text{leading to } F_{AH} = 88.5 \text{ (tension)}$$

## 102

The solution will assume that the critical condition is overturning due to wind gusts.

$$\text{Resultant wind force on the sign} = 14 \times 18 \times \frac{55}{144} = 96.25 \text{ lb}$$

Height to center of sign = 41 in.

Overturning moment about the tipping heel = 96.25 × 41 = 3946.3 lb-in.

Stabilizing moment = $W_{bucket}$ × 7 in. = 3946.3. Therefore, $W_{bucket}$ = 563.8 lb          **D**

## 103

Considering an 18-in.-long strip of the angle (load incident on ONE bolt), weight of the wall panel = 90 × 2/12 × 6 × 18/12 = 135 lb.

The resultant compression ($B_x$) on the wood blocking is considered to act at mid-height. Therefore, the distance between $A_x$ and $B_x$ is 5.5 in. The FBD of this section of the angle is shown below.

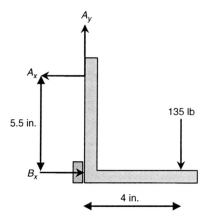

The axial tension in the bolt is represented by $A_x$.

Taking moments about B: $-5.5A_x + 135 \times 4 = 0 \Rightarrow A_x = 98.2$ lb          **B**

## 104

Using the stress transformation equations (2D), the maximum bending stress is given by:

$$\sigma_{max} = \frac{\sigma_{xx} + \sigma_{yy}}{2} + \sqrt{\left(\frac{\sigma_{xx} - \sigma_{yy}}{2}\right)^2 + \tau_{xy}^2} = \frac{28.5 + 0}{2} + \sqrt{\left(\frac{28.5 - 0}{2}\right)^2 + 14.8^2} = 34.8 \text{ ksi}$$

(Continued on next page)

The maximum shear stress is given by:

$$\sigma_{max} = \sqrt{\left(\frac{\sigma_{xx} - \sigma_{yy}}{2}\right)^2 + \tau_{xy}^2} = \sqrt{\left(\frac{28.5 - 0}{2}\right)^2 + 14.8^2} = 20.6 \text{ ksi}$$

Factor of safety based on normal stress: $FS = \dfrac{\sigma_{ult}}{\sigma_{max}} = \dfrac{50}{34.8} = 1.44$

Factor of safety based on shear stress: $FS = \dfrac{\tau_{ult}}{\tau_{max}} = \dfrac{30}{20.6} = 1.46$

Overall (lowest) factor of safety = 1.44      **A**

# 105

The critical design condition for the pier at C is the live load placement, which maximizes support reaction at pier C. The qualitative influence line diagram for the vertical reaction at C is shown below. Influence lines for statically determinate structures are composed of linear segments but for indeterminate structures, influence lines are curved.

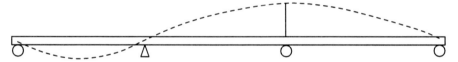

Therefore, to maximize vertical reaction at pier C, the live load must be placed on spans BC and CD. Correct answer is      **D**

# 106

The tributary width for each joist is the center to center distance = 3 ft

Floor load incident to each joist = 90 lb/ft² ×3 ft = 270 lb/ft

Maximum bending moment in simply supported joists: $M_{max} = \dfrac{wL^2}{8} = \dfrac{270 \times 18^2}{8} =$ 10,935 lb·ft = 131,220 lb·in.

The allowable bending stress = 1,700 psi

Required (minimum) section modulus: $S = \dfrac{M}{\sigma_{all}} = \dfrac{131,220}{1,700} = 77.2 \text{ in.}^3$
(Choose 80 in.³)      **C**

## 107

The free body diagram of joint B is shown below. Since this FBD has two unknowns, the two equations of equilibrium may be used to solve for them.

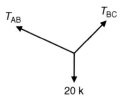

$$\sum F_x = -T_{AB}\frac{3}{\sqrt{10}} + T_{BC}\frac{1}{\sqrt{2}} = 0 \Rightarrow T_{AB} = 0.745T_{BC}$$

$$\sum F_y = T_{AB}\frac{1}{\sqrt{10}} + T_{BC}\frac{1}{\sqrt{2}} - 20 = 0$$

$$\Rightarrow \left(0.745\frac{1}{\sqrt{10}} + \frac{1}{\sqrt{2}}\right)T_{BC} = 20 \Rightarrow T_{BC} = 21.21\text{ k}, T_{AB} = 15.81\text{ k}$$

A

### Alternate solution

Since it has three forces in equilibrium, the force polygon is a triangle. Use the law of sines.

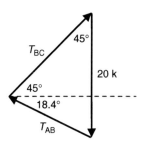

$$\frac{T_{AB}}{\sin 45} = \frac{20}{\sin 63.4} \Rightarrow T_{AB} = 15.82\text{ k}$$

## 108

- Since the load is vertical, the only horizontal reaction (at node I) is zero.
- At node I, there are three forces (ONLY TWO LINES OF ACTION) in equilibrium ($F_{HI}$, $F_{GI}$, and $I_y$), of which $F_{HI}$ acts alone along its line of action. *Therefore, $F_{HI}$ must be zero.*
- With $F_{HI}$ = zero, at node H, there are three forces (ONLY TWO LINES OF ACTION) in equilibrium ($F_{FH}$, $F_{HG}$, and $H_y$), of which $F_{HG}$ acts alone along its line of action. *Therefore, $F_{HG}$ must be zero.*
- With $F_{HG}$ = zero, at node G, there are three forces (ONLY TWO LINES OF ACTION) in equilibrium ($F_{EG}$, $F_{FG}$, and $F_{GI}$), of which $F_{FG}$ acts alone along its line of action. *Therefore, $F_{FG}$ must be zero.*
- With $F_{FG}$ = zero, at node F, there are three forces (ONLY TWO LINES OF ACTION) in equilibrium ($F_{DF}$, $F_{FE}$, and $F_{FH}$), of which $F_{FE}$ acts alone along its line of action. *Therefore, $F_{FE}$ must be zero.*
- With $F_{FE}$ = zero, at node E, there are three forces (ONLY TWO LINES OF ACTION) in equilibrium ($F_{DE}$, $F_{CE}$, and $F_{EG}$), of which $F_{DE}$ acts alone along its line of action. *Therefore, $F_{DE}$ must be zero.*

*Note:* Beyond this point, the condition of ONLY TWO LINES OF ACTION is not satisfied at any of the nodes A, B, C, D.

Zero force members are: HI, HG, FG, FE, DE                                               **C**

## 109

$F_{200} = 10$

Since this is less than 50, it is predominantly a coarse grained soil (first letter G or S). Also, since $F_{200}$ is between 5 and 12%, soil will have dual classification. This eliminates **A**.

Coarse fraction = 90%

$R_4 = 100 - 41 = 59$

This is more than half of the coarse fraction. Therefore, first letter is G.

Note, in these problems, solution strategy may depend heavily on given answer choices. All given choices have first letter G. Therefore, the previous step is redundant.

$D_{10}$ = No. 200 size = 0.075 mm

$D_{30}$ = No. 10 size = 2.0 mm (slightly less)

$D_{60}$ = 0.5 in. = 12.7 mm

$$C_u = \frac{D_{60}}{D_{10}} = \frac{12.5}{0.075} = 166.7 \qquad\qquad C_c = \frac{D_{30}^2}{D_{10}D_{60}} = \frac{2.0^2}{0.075 \times 12.5} = 4.3$$

Since $F_{200}$ is between 5 and 12%, the soil has a dual classification. The first part of the classification is based on gradation. Since *both* criteria ($C_u > 4$ and $1 < C_c < 3$) for GW are not met, the soil must be classified GP. This eliminates **D**.

GP-GM if PI < 0.73(LL–20) OR PI < 4

GP-GC if PI > 0.73(LL–20) AND PI > 7

PI = 54 – 23 = 31

Value of PI on the A-line = 0.73(LL–20) = 24.8. Soil plots above the A-line (clay). Thus, the soil meets BOTH criteria for GP-GC.  **B**

## 110

| Sieve Analysis | Granular Materials (35% or less passing no. 200 sieve) | | | | | | | Silt-Clay Materials (more than 35% passing no. 200 sieve) | | | | |
| --- | --- | --- | --- | --- | --- | --- | --- | --- | --- | --- | --- | --- |
| | A-1 | | A-3 | A-2 | | | | A-4 | A-5 | A-6 | A-7 | A-8 |
| | A-1-a | A-1-b | | A-2-4 | A-2-5 | A-2-6 | A-2-7 | | | | | |
| % passing | | | | | | | | | | | | |
| No. 10 | ≤50 | | | | | | | | | | | |
| No. 40 | ≤30 | ≤50 | >50 | | | | | | | | | |
| No. 200 | ≤15 | ≤25 | ≤10 | ≤35 | ≤35 | ≤35 | ≤35 | >35 | >35 | >35 | >35 | |
| LL | | | | ≤40 | >40 | ≤40 | >40 | ≤40 | >40 | ≤40 | >40 | |
| PI | ≤6 | | NP | ≤10 | ≤10 | >10 | >10 | ≤10 | ≤10 | >10 | >10 | |

Since $F_{200} < 35$, the groups A-4, A-5, A-6, and A-7 are eliminated. This eliminates choices **A** and **C**.

(*Continued on next page*)

Also, the soil has significant plasticity (PI = 31), so A-1 and A-3 are eliminated. These designations are essentially for soils with no to low plasticity. Given the four answer choices, this determination is redundant.

Soil meets criteria for A-2-7 ($F_{200}$ < 35, LL > 40, PI > 10). **B**

## 111

Ignoring friction between backfill and the wall (Rankine theory), the active earth pressure coefficient is given by (simplified equation applies for $\theta = 90$, $\delta = 0$, $\beta = 0$):

$$K_a = \frac{1 - \sin 31}{1 + \sin 31} = 0.32$$

Active earth pressure resultant is given by:

$$R_a = \frac{1}{2}K_a \gamma H^2 = 0.5 \times 0.32 \times 118.4 \times 15^2 = 4{,}262.4 \text{ lb/ft} \qquad \textbf{C}$$

## 112

The volume of the standard Proctor mold is 1/30 ft³. This volume is used to reduce the net soil mass to wet unit weight. For example, for sample 1: 3.2 lb ÷ (1/30) = 96 lb/ft³.

The total unit weight is then converted to a dry unit weight using $\gamma_d = \gamma/(1 + w) = 96/1.128 = 85.1$.

| Weight of Soil (lb) | Water Content (%) | Unit Weight (pcf) | Dry Unit Weight (pcf) |
|---|---|---|---|
| 3.20 | 12.8 | 96.0 | 85.1 |
| 3.78 | 13.9 | 113.4 | 99.6 |
| 4.40 | 15.0 | 132.0 | 114.8 |
| 4.10 | 15.7 | 123.0 | 106.3 |
| 3.70 | 16.6 | 111.0 | 95.2 |
| 3.30 | 18.1 | 99.0 | 83.9 |

The maximum dry unit weight = 114.8 lb/ft³ **D**

## 113

Total unit weight = 127 pcf

Assume a unit volume approach. $V = 1$ cu ft

Therefore, total weight $W = 127$ lb

Since the water content is 10%, distribute the total weight in 100:10 proportions (solids:water). Therefore, weight of solids is $100/110 \times 127 = 115.5$ lb and weight of water is 11.5 lb.

*Note:* A common mistake is to calculate weight of water = 10% of 127 = 12.7 lb

Volume of solids $= \dfrac{115.5}{165.4} = 0.698$ cu ft

Volume of voids $= 1.0 - 0.698 = 0.302$

Porosity: $n = \dfrac{V_{voids}}{V_{total}} = 0.302 = 30\%$

**D**

## 114

Weight of soil solids: $W_s = 3.67$ lb

Weight of water: $W_w = 4.18 - 3.67 = 0.51$ lb

Water content: $w = \dfrac{0.51}{3.67} = 0.139$

Total unit weight: $\gamma = \dfrac{4.18}{1/30} = 125.4$ pcf

Specific gravity: $G_s = 2.73$

Using the formula:

$$\gamma = \frac{(1 + w)\, SG_s}{w G_s + S}\gamma_w \Rightarrow 125.4 = \frac{1.139 \times 2.73 \times S}{0.139 \times 2.73 + S} \times 62.4 \Rightarrow S = 0.693$$

**A**

## 115

Head difference causing seepage flow = 945 − 905 = 40 ft

Length of seepage path = 160 ft

Hydraulic gradient, $i = 40/160 = 0.25$ ft/ft

Area (rectangular) of flow, $A = 120 \times 6.5 = 780$ sq. ft

Permeability, $K = 9.5$ ft/day = 0.0066 ft/min

By Darcy's law, discharge:

$$Q = KiA = 0.0066 \times 0.25 \times 780 = 1.286 \text{ ft}^3/\text{min} = 9.62 \text{ gal/min}$$   **B**

## 116

For sample A:  Normal stress $\sigma_1 = 1,000$ psf
Shear stress $\tau_1 = 675$ psf

For sample B:  Normal stress $\sigma_2 = 3,000$ psf
Shear stress $\tau_2 = 2,025$

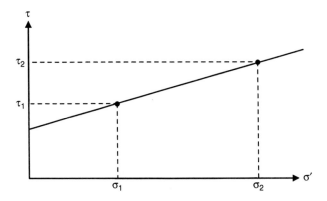

Friction angle is calculated from:  $\tan\phi = \dfrac{\tau_2 - \tau_1}{\sigma_2 - \sigma_1} = \dfrac{2,025 - 675}{3,000 - 1,000} = 0.675 \Rightarrow \phi = 34°$

Cohesion: $c = \dfrac{\tau_1\sigma_2 - \tau_2\sigma_1}{\sigma_2 - \sigma_1} = \dfrac{675 \times 3,000 - 2,025 \times 1,000}{3,000 - 1,000} = 0$   **A**

## 117

Wastewater flow rate = 750 gpm = 750/448.8 = 1.67 ft³/sec

Immediately downstream of mixing location, dissolved oxygen concentration is calculated using the weighted average of the two mixing streams:

$$\overline{DO} = \frac{18 \times 5.1 + 1.67 \times 1.7}{18 + 1.67} = 4.81 \text{ mg/L}$$

B

## 118

The total solids load in the influent: $X = 3 \times 180 \times 8.3454 = 4{,}506.5$ lb/day

Number of units needed (if all are operational): $n = 4{,}506.5/800 = 5.6$ (6 filters)

With the requirement of one offline printer at a time, 7 filters are needed.

C

## 119

The 1-hr unit hydrograph is used to construct the runoff contribution of the first hour and the second hour (staggered).

| Time (hr) | 0 | 1 | 2 | 3 | 4 | 5 |
|---|---|---|---|---|---|---|
| Discharge Q (cfs/in.) | 0 | 35 | 75 | 105 | 40 | 0 |

| Time (hr) | 0 | 1 | 2 | 3 | 4 | 5 | 6 |
|---|---|---|---|---|---|---|---|
| Hour 1 (scaled × 1.7) | 0 | 59.5 | 127.5 | 178.5 | 68 | 0 | |
| Hour 2 (scaled × 0.8) | | 0 | 28 | 60 | 84 | 32 | 0 |
| Total | 0 | 59.5 | 155.5 | 238.5 | 152 | 32 | 0 |

Peak discharge = 238.5 cfs

B

## 120

The longitudinal slope ($S$) and the channel roughness ($n$) *do not* determine the *critical* flow parameters (such as depth or velocity). Therefore, for the question asked, they are unnecessary. However, they would be required data for the calculation of the *normal* depth.

For a triangular channel with side slope parameter $m$, the critical depth is given by:

$$d_c = \left(\frac{2Q^2}{gm^2}\right)^{1/5} = \left(\frac{2 \times 10^2}{32.2 \times 3^2}\right)^{1/5} = 0.93 \text{ ft}$$

Critical velocity is given by: $V_c = \left(\frac{Qg^2}{4m}\right)^{1/5} = \left(\frac{10 \times 32.2^2}{4 \times 3}\right)^{1/5} = 3.87 \text{ ft/sec}$       **C**

These relations are derived from the general expression: $Q^2 T = A^3 g$ applied to the triangular section shown below:

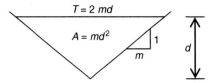

$$(Q^2)(2md_c) = (md_c^2)^3 g \Rightarrow d_c = \left(\frac{2Q^2}{gm^2}\right)^{1/5}$$

Critical velocity: $V_c = \dfrac{Q}{A} = \dfrac{Q}{md_c^2} = \left(\dfrac{Qg^2}{4m}\right)^{1/5}$

## 121

Longitudinal slope, $S = 0.005$

Side slope parameter, $m = 3$

Manning's roughness coefficient, $n = 0.020$

Bottom width, $b = 2$ ft

Depth of flow, $d = 2$ ft

Therefore, horizontal flare (each side) = $3 \times 2 = 6$ ft

Inclined flare (each side) = $\sqrt{2^2 + 6^2}$ √(2 = $\sqrt{40}$ √40 = 6.325 ft

Wetted perimeter, $P = 2 + 2\sqrt{40} = 14.65$ ft

Top width = 14 ft

Flow area, $A = \dfrac{2+14}{2} \times 2 = 16$ ft$^2$

Hydraulic radius = $16 \div 14.65 = 1.09$ ft

$V = \dfrac{1.486}{0.02} \times 1.09^{2/3} \times 0.005^{1/2} = 5.56$ ft/sec          A

## Alternate solution

Using Table 303.3 (in I. Goswami, *Civil Engineering PE All-in-One Exam Guide: Breadth and Depth*, McGraw-Hill, 2009), for $d/b = 2/2 = 1.0$ and $m = 3$, the parameter $K = 2.6725$. Using this parameter value in Equation 303.36:

Flow rate, $Q = \dfrac{2.6725 \times 1.486 \times 2^{8/3} \times 0.005^{1/2}}{0.020} = 89.15$ cfs

Flow velocity, $V = \dfrac{Q}{A} = \dfrac{89.15}{16} = 5.57$ ft/sec

This would require calculation of the flow area only.

# 122

The suspended solids are in the fraction that is retained on the filter paper. The dissolved solids are in the fraction that passes through the filter paper.

The volatile dissolved solids concentration (mg/L) of the water sample is calculated as the difference between:

Mass of dry evaporation dish + solids = 275.801 g

Mass of evaporation dish + ignited solids = 275.645 g

Volatile solids in a 200-mL sample = 275.801 − 275.645 = 0.156 g

VDS concentration = 0.156 g/200 mL = 780 mg/L          C

## 123

The slope of the pipe is 0.01. Height difference between the ends of the pipe = 125 − 95 = 30 ft. Therefore, the length of the pipe = 30/0.01 = 3,000 ft

Hazen Williams, C = 120

If flow velocity is V (ft/sec), head loss due to friction is given by:

$$h_f = \frac{3.022 \times V^{1.85} \times 3,000}{120^{1.85} \times 1.0^{1.165}} = 1.291 V^{1.85}$$

Bernoulli's equation applied between free surface at reservoir and free flow at the outfall:

$$145 + \frac{p_{atm}}{\gamma} + 0 - 1.29 V^{1.85} = 95 + \frac{p_{atm}}{\gamma} + \frac{V^2}{2g} \Rightarrow 0.0155 V^2 + 1.29 V^{1.85} = 50$$

Solving approximately (trial and error): V = 7.15 ft/sec

Flow rate, $Q = 7.15 \times \frac{\pi}{4} \times 1^2 = 5.62$ cfs $= 2,516$ gal/min                     D

### Alternate solution

If the $V^2$ term is approximated by $V^{1.85}$, V = 7.17 ft/sec (without trial and error).

## 124

The question directs us to use the Rational Method, even though it is usually not used for areas larger than about 100 acres.

Time of concentration = longest of all overland flow times = 45 min

From the Intensity-Duration-Frequency curves, for a 20-year storm, with duration = 45 min, we get intensity I = 3.7 in./hr.

The composite Rational C coefficient is given by:

$$\bar{C} = \frac{\sum C_i A_i}{\sum A_i} = \frac{0.4 \times 80 + 0.2 \times 80 + 0.9 \times 50 + 0.6 \times 90 + 0.2 \times 70}{370} = 0.44$$

Rational method runoff discharge:

$$Q = CiA = 0.44 \times 3.7 \times 370 = 603 \text{ ac} \cdot \text{in./hr} = 608 \text{ ft}^3/\text{sec}$$ B

## 125

The tangent offset at any location on a vertical curve $= \frac{1}{2} Rx^2$

At the end of the curve (i.e., at the PVT), $x = L$, therefore:

$$\text{tangent offset} = \frac{1}{2}\frac{G_2 - G_1}{L}L^2 = \frac{(G_2 - G_1)L}{2}$$

For a crest curve, the vertical offset is negative, therefore:

$$\frac{(-4-5)L}{2} = -17.65 \Rightarrow L = 3.9222 \text{ sta}$$

Therefore, since the PVC is half the curve length upstream of the PVI:

Sta. PVC = sta. PVI − 1.9611 = 123.325 − 1.9611 = 121.3639 ft (sta. 121 + 36.39)   D

## 126

Total number of observations = 88

85% of observations = 74.8

Speed of 45 mph has an associated cumulative frequency = 74

85th percentile speed should be marginally higher than 45 mph (interpolation gives 45.4).   C

## 127

Question asks *only* for *braking distance*. The default value (AASHTO Green Book) of 11.2 ft/s² is assumed for the deceleration rate. Braking distance is given by:

$$d_b = \frac{1.075 \times S_{mph}^2}{(a \pm gG)} = \frac{1.075 \times 40^2}{11.2 + 32.2 \times 0.02} = 145 \text{ ft}$$   A

## 128

For acceleration phase, time = 70/8 = 8.75 sec

Acceleration rate = 8 mph/sec = 8 × 1.47 = 11.76 ft/s²

Distance traveled during acceleration phase = ½ at² = 0.5 × 1.76 × 8.75² = 450.2 ft

For deceleration phase, time = 70/10 = 7.0 sec

Deceleration rate = 10 mph/sec = 10 × 1.47 = 14.7 ft/s²

Distance traveled during acceleration phase = ½ at² = 0.5 × 14.7 × 7² = 360.2 ft

Total distance = 0.5 mile = 2,640 ft

Distance for constant velocity phase = 2,640 − 450.2 − 360.2 = 1,829.6 ft

During this phase, speed = 70 × 1.47 = 102.9 ft/sec

Time for constant velocity phase = 1,829.6/102.9 = 17.8 sec

Total travel time = 8.75 + 17.8 + 7.0 = 33.53 sec

Average running speed = 2,640/33.53 = 78.73 fps = 53.6 mph          **C**

## 129

For the given conditions, $e + f = 0.12 + 0.08 = 0.20$

$$\frac{V^2}{15R} \le e + f = 0.20 \Rightarrow R \ge \frac{60^2}{15 \times 0.02} = 1,200 \text{ ft}$$

When the intersecting roads are perpendicular, $I = 90°$, minimum length of curve:

$$L = \frac{RI}{57.29578} = \frac{1,200 \times 90}{57.29578} = 1,885 \text{ ft}$$          **A**

## 130

The radius of the curve is: $R = \dfrac{5,729.578}{D} = \dfrac{5,729.578}{4} = 1,432.4$ ft

The tangent length is the distance from PC to PI:

$$T = \sqrt{(1,232.56 - 509.72)^2 + (123.32 + 172.11)^2} = 780.88 \text{ ft}$$

The azimuth of the back tangent is calculated from:

$$\tan Az = \frac{\Delta E}{\Delta N} = \frac{-172.11 - 123.32}{509.72 - 1{,}232.56} \Rightarrow Az = 202.23° \text{ (third quadrant angle because}$$
both $\Delta E$ and $\Delta N$ are negative)

Deflection angle: $I = 2\tan^{-1}\left(\dfrac{T}{R}\right) = 57.2°$ to the left (counterclockwise)

Therefore, azimuth of forward tangent = $202.23 - 57.2 = 145.03$

Change in coordinates from PI to PT (using tangent length $T = 780.88$ ft)

$$\Delta N = T\cos Az = 780.88 \times \cos 145.03 = -639.90 \text{ ft}$$

$$\Delta E = T\sin Az = 780.88 \times \sin 145.03 = +447.56 \text{ ft}$$

Coordinates of the PT: $(509.72 - 639.90, -172.11 + 447.56)$ or
(130.18 S, 275.45 E)                                                          D

# 131

Algebraic difference of grades: $A = |G_2 - G_1| = 3.6\%$

Design speed = 65 mph

Corresponding stopping sight distance (AASHTO Green Book) = $239 + 405 = 644$ ft

Assuming the condition $S < L$, the length of curve for adequate stopping sight distance (AASHTO Green Book) is given by:

$$L = \frac{AS^2}{2{,}158} = \frac{3.6 \times 644^2}{2{,}158} = 692 \text{ ft}$$

This solution fits the criterion $S < L$. Therefore, the solution is OK.

Minimum length of curve = 692 ft                                              B

## 132

Pedestrians using the walkway in the first hour $= 0.9 \times 0.3 \times 40{,}000 = 10{,}800$ ped/hr (hourly average)

Peak flow during the first hour $= 10{,}800/0.88 = 12{,}273$ ped/hr $= 204.5$ ped/min

Peak flow rate $= 204.5/32 = 6.4$ ped/min/ft                               C

## 133

The daily production rate (yd$^3$) in an 8-hr workday is most nearly:

In 8 hr, number of cycles $= 480$ min/8 min $= 60$

Ideal daily production $= 3 \times 60 = 180$ yd$^3$

Actual production $= 0.90 \times 0.72 \times 180 = 116.64$ yd$^3$               A

## 134

A has $ES_A = 0$, $EF_A = 4$

D has (single predecessor A) $ES_D = 4$, $EF_D = 11$

B has $ES_B = 0$, $EF_B = 3$

E has (single predecessor B) $ES_E = 3$, $EF_E = 9$

G has $ES_G = $ larger of $EF_D = 11$ and $EF_E = 9$

Therefore, $ESG = 11$ weeks                               C

## 135

Assume 1 sack cement:

Weight of cement, $W_c = 94$ lb

Volume of cement, $V_c = \dfrac{94}{3.15 \times 62.4} = 0.478$ ft$^3$

SSD sand: Weight, $W_s = 1.6 \times 94 = 150.4$ lb

Volume of SSD sand, $V_s = \dfrac{150.4}{2.62 \times 62.4} = 0.920$ ft$^3$

SSD coarse aggregate: Weight, $W_{CA} = 2.6 \times 94 = 244.4$ lb

Volume of coarse aggregate, $V_{CA} = \dfrac{244.4}{2.65 \times 62.4} = 1.478$ ft$^3$

Water: Volume of water, $V_w = 5.8$ gal $= 0.775$ ft$^3$

Weight of water, $W_w = 0.775 \times 62.4 = 48.4$ lb

The components above (V = 3.651 ft$^3$) represent 97% of total volume because air = 3%. Therefore, total volume = 3.651/0.97 = 3.764 ft$^3$

The quantity of cement = 94 lb/3.764 ft$^3$ = 674 lb/yd$^3$      A

## 136

Truckload = 10 tons = 20,000 lb of soil

Soil volume in each truckload $= \dfrac{20,000}{125} = 160$ ft$^3$ = 5.926 yd$^3$

Thus, truck moves 5.926 yd$^3$ every 30 min.

So, in a 10-hr workday, a single truck makes 600 min/30 min = 20 trips, moving 20 × 5.926 = 118.52 yd$^3$. In 8 days, a single truck moves 948.2 yd$^3$.

Total number of trucks needed = 4,200/948.2 = 4.43

Use 5 trucks.      B

## 137

Excavation for grade beam has a bottom width of 3 ft, 1:1 side slopes, and a depth of 2 ft, which creates a top width:

$T = 3 + 2 \times 2 = 7$ ft

Cross section of trench: $A = \dfrac{1}{2} \times (3+7) \times 2 = 10$ ft$^2$

(*Continued on next page*)

Perimeter of building: $P = 200 \times 2 + 400 \times 2 + 50 \times 2 = 1,300$ ft

Volume of excavation: $V = 10 \times 1,300 = 13,000$ ft$^3 = 481.5$ yd$^3$

Daily productivity $= 9$ yd$^3$/hr $\times 8$ hr/day $= 72$ yd$^3$/day

Number of days $= 481.5/72 = 6.7$ days

Choose 7 days as minimum number of days to complete job.      **D**

## 138

One set of forms will be new (area of both sides $= 2 \times 12 \times 20 = 480$ ft$^2$) and two sets (area $= 960$ ft$^2$) will be reused.

Volume of concrete, not including waste: $V_c = 12 \times 60 \times 1 = 720$ ft$^3 = 26.67$ yd$^3$

Total volume of concrete, including waste: $V_c = 1.1 \times 26.67 = 29.33$ yd$^3$

Therefore, cost of erecting forms $= 4.30 \times 480 + 1.30 \times 960 = \$3,312$

Cost of dismantling forms $= 1.05 \times 1,440 = \$1,512$

Cost of concrete (including waste) $= 120 \times 29.33 = \$3,519.60$

Cost of reinforcement (not including waste) $= 25 \times 26.67 = \$666.67$

Total cost $= \$9,011$      **D**

## 139

Using trapezoidal method, the volume of excavation is calculated using:

$$V = \frac{\Delta}{2}\left[ y_0 + y_n + 2\sum_1^{n-1} y_i \right]$$

$$V = \frac{50}{2} \times [456.33 + 493.34 + 2 \times (563.97 + 702.24 + 1,234.98 + 783.92 + 591.94)]$$

$$= 217,594.25 \text{ ft}^3 = 8,059 \text{ yd}^3 \qquad \textbf{A}$$

## 140

| | | |
|----|----------|-------|
| AD | 11 weeks | 8,000 |
| AE | 6 weeks  | 6,000 |
| BE | 7 weeks  | 7,000 |
| CF | 10 weeks | 9,000 |

To complete project in 9 weeks, AD must be shortened by 2 weeks (shorten D by 2 weeks for additional $1,000) AND CF must be shortened by 1 week (shorten C by 2 weeks for additional $3,000). Bonus due to 2 weeks early completion is $2,000.

Net revised cost = original cost + extra cost − bonus = 24k + 1k + 3k − 2k = $26,000

D

## Answer Key for Breadth Exam No. 2

| | | | | | | | | | |
|---|---|---|---|---|---|---|---|---|---|
| 101 | B | 111 | C | 121 | A | 131 | B |
| 102 | D | 112 | D | 122 | C | 132 | C |
| 103 | B | 113 | D | 123 | D | 133 | A |
| 104 | A | 114 | A | 124 | B | 134 | C |
| 105 | D | 115 | B | 125 | D | 135 | A |
| 106 | C | 116 | A | 126 | C | 136 | B |
| 107 | A | 117 | B | 127 | A | 137 | D |
| 108 | C | 118 | C | 128 | C | 138 | D |
| 109 | B | 119 | B | 129 | A | 139 | A |
| 110 | B | 120 | C | 130 | D | 140 | D |

# 10

# Structural Depth Exam Solutions

These detailed solutions are for questions 201 to 240, representative of a 4-hour structural depth exam according to the syllabus and guidelines for the Principles and Practice (P&P) of Civil Engineering Examination, administered by the National Council of Examiners for Engineering and Surveying (NCEES).

## 201

Self weight of the beam is given by:

$$w_{self} = \frac{15 \times 28}{144} \times 0.15 = 0.44 \text{ k/ft}$$

Factored load is calculated as:

$$w_u = 1.2w_D + 1.6w_L = 1.2 \times (0.44 + 2.4) + 1.6 \times 4 = 9.81 \text{ k/ft}$$

Effective span = Center to center between bearings = 24 ft 10 in. = 24.83 ft

Design moment, $M_{max} = \dfrac{wL^2}{8} = \dfrac{9.81 \times 24.83^2}{8} = 756 \text{ k} \cdot \text{ft}$

*(Continued on next page)*

If ultimate moment is expressed as $M_u = \phi f'_c bd^2 X$, then the strength parameter X is given by:

$$X = \frac{M_u}{\phi f'_c bd^2} = \frac{756 \times 12}{0.9 \times 5 \times 15 \times 25.5^2} = 0.207$$

From design tables (Table 105.5 in I. Goswami, *Civil Engineering PE All-in-One Exam Guide: Breadth and Depth*, McGraw-Hill, 2009):

$$w = \frac{\rho f_y}{f'_c} = 0.242 \Rightarrow \rho = \frac{w f'_c}{f_y} = \frac{0.242 \times 5}{60} = 0.02$$

Area of steel required: $A_s = \rho bd = 0.02 \times 15 \times 25.5 = 7.65$ in.$^2$    **D**

## 202

Self weight of the beam is given by:

$$w_{self} = \frac{15 \times 28}{144} \times 0.15 = 0.44 \text{ k/ft}$$

Factored load is calculated as:

$$w_u = 1.2 w_D + 1.6 w_L = 1.2 \times (0.44 + 2.4) + 1.6 \times 4 = 9.81 \text{ k/ft}$$

Effective span = Center to center between bearings = 24 ft 10 in. = 24.83 ft

Maximum support reaction (center of bearing): $V_u = \dfrac{w_u L}{2} = \dfrac{9.81 \times 24 \cdot 83}{2} = 121.8 \text{ k}$

Since the effective depth of the beam is 25.5 in., design shear at critical section [which is 25.5 in. from face of support, therefore 30.5 in. (2.5 ft) from center of bearing] = 121.8 − 9.81 × 2.5 = 97.3 k.

Shear capacity of unreinforced section is calculated as:

$$V_c = 2\sqrt{f'_c} bd = 2 \times \sqrt{5,000} \times 15 \times 25.5 = 54,094 \text{ lb}$$

Required shear capacity of shear reinforcement is given by:

$$V_s = \frac{V_u}{\phi} - V_c = \frac{97.3}{0.75} - 54.1 = 75.6 \text{ k}$$

Maximum permitted spacing is given by:

$$s = \frac{A_v f_y d}{V_s} = \frac{0.4 \times 60 \times 25.5}{75.6} = 8.1 \text{ in.}$$

Choose 7 in.                                                                C

## 203

Since the column to beam connections are simple (moment free), the maximum moment for the beam (BC) is:

$$M_u = \frac{w_u L^2}{8} = \frac{8 \times 30^2}{8} = 900 \text{ k} \cdot \text{ft}$$

From moment capacity charts (AISC Steel Manual, 13th edition), for $M_u = 900$ k-ft and $L_b = 30$ ft, and $F_y = 50$ ksi, lightest section is W24 × 131.              B

## 204

### ASD Solution

Required available strength is calculated as: $P_a = 200 + 200 = 400$ k

The W12 × 96 section has the following properties: $r_x = 5.44$; $r_y = 3.09$

Slenderness ratios are calculated as: $(KL/r)_x = 88.2$; $(KL/r)_y = 77.7$

From table 4-22 of the AISC Steel Manual, for the governing slenderness ratio = 88.2, allowable stress is given by: $F_{cr}/\Omega_c = 14.28$ ksi.

Allowable compressive load is given by: $P_n/\Omega_c = 14.28 \times 28.2 = 402.7$ k

### LRFD Solution

$$P_u = 1.2 \times 200 + 1.6 \times 200 = 560 \text{ k}$$

The W12 × 96 section has the following properties: $r_x = 5.44$; $r_y = 3.09$

Slenderness ratios are calculated as: $(KL/r)_x = 88.2$; $(KL/r)_y = 77.7$

(Continued on next page)

From table 4-22 of the AISC Steel Manual, for the governing slenderness ratio = 88.2, design buckling stress is given by: $\phi_c F_{cr} = 21.56$ ksi.

Design strength in compression is given by: $\phi_c P_n = 21.56 \times 28.2 = 608$ k

Check the next lower size.

W12 × 87: $r_x = 5.38$; $r_y = 3.07$; $(KL/r)_x = 89.2$; $(KL/r)_y = 78.2$; $\phi_c F_{cr} = 21.36$ ksi; $\phi_c P_n = 21.36 \times 25.6 = 547$ k (inadequate)

Choose W12 × 96                                                                                          C

## 205

The effective span of the lintel is calculated as S = center to center distance between bearings = 5.67 ft.

Height of wall above the lintel = 8 ft > 5.67 ft. Therefore, we can assume that arching action occurs.

Thus, the lintel needs to carry two kinds of loading:

1. self-weight (rectangular): $w_1 = 140 \times \dfrac{7.5}{12} \times \dfrac{16}{12} = 116.7$ lb/ft

2. wall weight (triangular): $w_2 = 130 \times \dfrac{7.5}{12} \times \dfrac{5.67}{2} = 230.3$ lb/ft

Maximum bending moment in the simply supported lintel is given by

$$M_{max} = M_1 + M_2 = \frac{w_1 L^2}{8} + \frac{w_2 L^2}{12} = 469.0 + 617.0 = 1{,}086 \text{ lb} \cdot \text{ft}$$

Correct answer is                                                                                        D

## 206

Using the unit load method, the deflection at a point is given by the integral:

$$\delta = \int \frac{Mm}{EI} dx$$

M is the bending moment function under actual load and m is the bending moment function under virtual load. These are shown in the following figures.

Flexural rigidity $EI = 29000 \times 870 = 2.523 \times 10^7$ k-in.$^2 = 1.75 \times 10^5$ k-ft$^2$

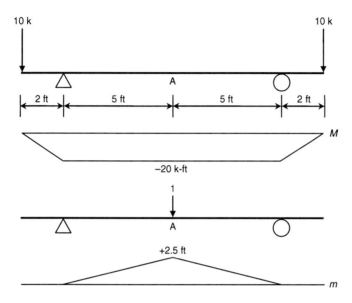

$$\delta = 2 \times \left[ \frac{1}{2} \times \frac{-20 \times 2.5 \times 5}{EI} \right] = -\frac{250}{EI} = -\frac{250}{1.75 \times 10^5} = -0.001429 \text{ ft} = -0.017 \text{ in.}$$

In the step above, the integral is facilitated by the use of Table 102.1 in *All-in-One*. The negative sign indicates that the deflection is opposite to the direction of the unit load (which was applied downward).                                                    C

## 207

According to AASHTO, effective width of slab = c/c distance between beams = 8 ft = 96 in.

Modular ratio: $n = \dfrac{E_s}{E_c} = \dfrac{29000 \text{ ksi}}{1820\sqrt{4} \text{ ksi}} = 7.97 = 8$

Equivalent width of slab = $\dfrac{96}{8} = 12$ in.

Height of centroid (measured from bottom of steel):

$$\bar{y} = \frac{136 \times 30 + 12 \times 8 \times 64}{136 + 96} = 44.1 \text{ in.}$$

*(Continued on next page)*

Moment of inertia of equivalent section:

$$I_{NA} = 81,940 + 136 \times (44.1 - 30)^2 + \frac{1}{12} \times 12 \times 8^3 + 96 \times (44.1 - 64)^2 = 147,507 \text{ in.}^4$$

Maximum bending moment in composite section:

$$M = \frac{wL^2}{8} = \frac{1.8 \times 8 \times 70^2}{8} = 8,820 \text{ k} \cdot \text{ft} = 105,840 \text{ k} \cdot \text{in.}$$

Bending stress in steel (tensile): $\sigma = \dfrac{My}{I} = \dfrac{105,840 \times 44 \cdot 1}{147,507} = 31.6 \text{ ksi}$   **A**

## 208

A572 grade 50 steel: $F_y = 50$ ksi; $F_u = 65$ ksi

For the channel (C10 × 30), relevant properties: $A_g = 8.81$ in.$^2$; $t_w = 0.673$ in.

Net area (2 holes): $A_{net} = 8.81 - 2 \times \dfrac{7}{8} \times 0.673 = 7.63$ in.$^2$

Net area (3 holes): $A_{net} = 8.81 - 3 \times \dfrac{7}{8} \times 0.673 + 2 \times \dfrac{2^2}{4 \times 2.5} \times 0.673 = 7.58$ in.$^2$

Shear lag factor: $U = 1 - \dfrac{\bar{x}}{L} = 1 - \dfrac{0.649}{4} = 0.84$

### ASD Solution

Capacity (yield): $\dfrac{P_n}{\Omega_t} = \dfrac{F_y A_g}{1.67} = \dfrac{50 \times 8.81}{1.67} = 263.8 \text{ k}$

Capacity (fracture): $\dfrac{P_n}{\Omega_t} = \dfrac{F_u U A_{net}}{2.0} = \dfrac{65 \times 0.84 \times 7.58}{2.0} = 206.9 \text{ k}$

Design member capacity = 207 k   **A**

### LRFD Solution

Capacity (yield): $\phi_t P_n = 0.9 F_y A_g = 0.9 \times 50 \times 8.81 = 396.5 \text{ k}$

Capacity (fracture): $\phi_t P_n = 0.75 F_u A_e = 0.75 F_u U A_{net} = 0.75 \times 65 \times 0.84 \times 7.58 = 310.4 \text{ k}$

Design member capacity = 310 k   **A**

## 209

Eccentricity is greater than 0.1h (4 in. > 0.1 × 16 in.). Therefore, the column must be designed for a combination of $P_u$ and $M_u$ ($e/h$ = 4/16 = 0.25)

Area of steel, $A_s$ = 12 × 1.0 = 12.0 in.², and gross area, $A_g$ = 16 × 16 = 256 in.²

Reinforcement ratio $\rho_g$ = 12/256 = 0.047 (This is within limits 1%–8%)

Assuming clear cover = 1.5 in., the center to center distance between parallel lines of reinforcement = 16 − 2 × (1.5 + 0.5 + 1.128/2) = 10.87 in.

Parameter $\gamma$ = 10.87/16 = 0.68

Let us use the diagram for $f'_c$ = 4 ksi, $f_y$ = 60 ksi, $\gamma$ = 0.65

From the chosen column interaction diagram, for $e/h$ = 0.25 and $\rho_g$ = 4.7%, $K_n$ = 0.80

$$P_u = K_n \phi_c f'_c A_g = 0.80 \times 0.65 \times 4 \times 256 = 532.5 \text{ k} \qquad \text{D}$$

## 210

The propped cantilever has degree of indeterminacy = 1. Therefore, for a plastic hinge collapse mechanism to form, *two* plastic hinges must form. These will occur at the locations of maximum moment, that is, at the support and at the point where the point load $P$ acts. At this condition, the collapse mechanism will be as shown below. Assuming rotations $\theta_1$ and $\theta_2$ at A and C, the compatibility equation for the deflection at B is: $12\theta_1 = 3\theta_2$

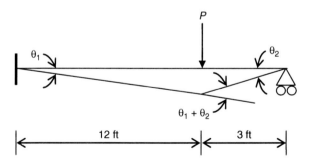

(Continued on next page)

According to the Principle of Virtual Work, work done by internal moments acting through deformations = work done by external loads acting through beam displacements:

$$M_p\theta_1 + M_p(\theta_1 + \theta_2) = P\delta$$

$$M_p\theta_1 + M_p(\theta_1 + 4\theta_1) = P \cdot 12\theta_1 \Rightarrow M_p = 2P$$

For W18 × 90 with $F_y = 50$ ksi, plastic moment capacity, $M_p = Z_x F_y = 186 \times 50 = 9300$ k-in. = 775 k-ft.

$$P = \frac{M_p}{2} = 387.5 \text{ k}$$ 

**C**

# 211

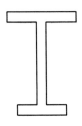

Top flange width = 12 in.
Bottom flange width = 10 in.
Top flange thickness = 1.5 in.
Bottom flange thickness = 2 in.
Overall depth = 22 in.
Web thickness = ¾ in.

Depth of web = 22 − 1.5 − 2.0 = 18.5 in.

The total section area, $A = 10 \times 2 + 0.75 \times 18.5 + 12 \times 1.5 = 51.875$ in.$^2$

Half area = 25.94 in.$^2$

Area of bottom flange = 20 in.$^2$. Therefore, web area below PNA = 25.94 − 20 = 5.94 in.$^2$ Depth of web below PNA = 5.94/0.75 = 7.917 in. This means the PNA is at a distance = 9.917 in. above bottom edge.

The plastic section modulus $Z_x$ is calculated as the first moment of the four component rectangles about the PNA:

$$Z_x = 20 \times 8.917 + 5.94 \times 3.958 + 7.94 \times 5.292 + 18 \times 11.333 = 447.86 \text{ in.}^3$$

Plastic moment capacity: $M_p = Z_x F_y = 447.86 \times 36 = 16,123 \text{ k} \cdot \text{in.} = 1,343.6 \text{ k} \cdot \text{ft}$   **A**

## 212

Centroid of axle system (distance from wheel 1): $\bar{x} = \dfrac{28 \times 10 + 30 \times 25}{68} = 15.15 \text{ ft}$

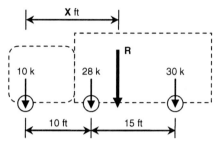

Position wheels such that wheel 2 (@10 ft) and resultant R (@ 15.15 ft) are equidistant from beam midspan ($x = 35$ ft). Distance between wheel 2 and resultant = $15.15 - 10 = 5.15$ ft

Therefore, wheel 2 is at $x = 35 - 5.15/2 = 32.43$ ft, wheel 1 is at 22.43 ft, and wheel 3 is at 47.43 ft.

The reaction at left support, due to this load placement, is calculated as:

$$R_A = \frac{47.57}{70} \times 10 + \frac{37.57}{70} \times 28 + \frac{22.57}{70} \times 30 = 31.5 \text{ k}$$

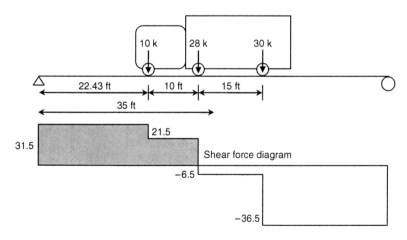

Maximum moment is calculated as the shaded area on the shear diagram:

$$M_{max} = 31.5 \times 22.43 + 21.5 \times 10 = 921.5 \text{ k} \cdot \text{ft}$$

C

## 213

Slab self weight: $w_{sw} = 150 \times \dfrac{5}{12} = 62.5 \text{ lb/ft}^2$

Total factored load on slab: $w_u = 1.2 \times (40 + 62.5) + 1.6 \times 85 = 259 \text{ lb/ft}^2$

For a unit width (1 ft) of the slab, the factored moment at the positive moment critical section is:

$$M_u = \frac{w_u L^2}{10} = \frac{259 \times 8^2}{10} = 1{,}657.6 \text{ lb} \cdot \text{ft} = 19{,}891.2 \text{ lb} \cdot \text{in.}$$

Assuming #5 main reinforcement bars and 0.75-in. clear cover, we get:

Effective depth of slab $= 5.0 - 0.75 - \dfrac{0.625}{2} = 3.94 \text{ in.}$

If ultimate moment is expressed as: $M_u = \phi f_c' b d^2 X$, then the strength parameter $X$ is given by:

$$X = \frac{M_u}{\phi f_c' b d^2} = \frac{19{,}891.2}{0.9 \times 4{,}500 \times 12 \times 3.94^2} = 0.0264 \Rightarrow w = \frac{\rho f_y}{f_c'} = 0.027$$

Reinforcement ratio, $\rho = \dfrac{w f_c'}{f_y} = 0.002$

As this reinforcement is less than the minimum required reinforcement, it must be increased by 33%.

Required area of steel, $A_s = \rho b d = 1.33 \times 0.002 \times 12 \times 3.94 = 0.126 \text{ in.}^2/\text{ft}$   **C**

## 214

Initial prestress (at release) $= 0.75 f_{pu} = 202.5 \text{ ksi}$

Prestressing force (after losses), $P_s = (202.5 - 32) \times 5.9 = 1{,}006 \text{ k}$

Eccentricity of the prestress force $= 21.65 \text{ in.}$

The moment due to prestress force induces tension on the top fiber, while the moment due to gravity loads induces compression on top.

$M_{DL+LL} = 1{,}400 \text{ k-ft} = 16{,}800 \text{ k-in.}$

Stress on top fiber (compression positive) is calculated as:

$$\sigma_{top} = +\frac{P_s}{A} - \frac{P_s e}{S_t} + \frac{M}{S_t} = +\frac{1{,}006}{283.8} - \frac{1{,}006 \times 21.65}{2{,}560} + \frac{16{,}800}{2{,}560} = +1.6 \text{ ksi (compression)} \quad \mathbf{B}$$

## 215

In order for block shear rupture to occur, four blocks must rupture (two in each flange) in order for the connection to fail. One of these blocks is shown below.

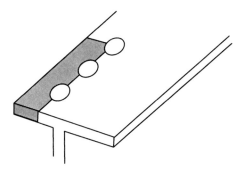

The block parameters (areas in tension and shear) are calculated:

$$A_{nt} = \left(3 - \frac{1}{2} \times \frac{7}{8}\right) \times 0.67 = 1.717 \text{ in.}^2$$

$$A_{gv} = 12 \times 0.67 = 8.04 \text{ in.}^2$$

$$A_{nv} = \left(12 - \frac{5}{2} \times \frac{7}{8}\right) \times 0.67 = 6.57 \text{ in.}^2$$

$$R_n = 0.6 F_u A_{nv} + U_{bs} F_u A_{nt} \le 0.6 F_y A_{gv} + U_{bs} F_u A_{nt}$$

$$R_n = 328.5 \le 273.3$$

Thus, the governing nominal strength of one block = 273.3 k. This is the nominal strength of one (of four) blocks. Thus, nominal strength in block shear = $4 \times 273.3 = 1{,}093.2$ k. $\qquad$ **B**

*Note:* Nominal and not design capacity is asked for in this question.

## 216

The Dead load + Wind load case produces maximum uplift.

Total dead load, $4 \times 150 + 60 = 660$ k, which produces compression in each pile = $660/27 = 24.4$ k

Wind load overturning moment is calculated as the sum of the moments of lateral forces about the top of the pile cap.

$M_{OT} = 30 \times 15 + 60 \times 30 + 70 \times 45 + 80 \times 60 = 10,200$ k · ft

Assuming linear force distribution in the columns (spacing $S$) and taking moments about pile line 5, we have 6 piles (lines 4 and 6) at distance $S$, 6 piles (lines 3 and 7) at distance $2S$, 6 piles (lines 2 and 8) at distance $3S$, and 6 piles (lines 1 and 9) at distance $4S$.

The sum of these moments = $6 \times F \times S + 6 \times 2F \times 2S + 6 \times 3F \times 3S + 6 \times 4F \times 4S = 180FS = 10,200$.

$F = 10,200/180 \times 6 = 9.44$ k

Thus, each pile in the outermost line of piles experiences a wind-uplift = $4F = 37.78$ k.

Gross uplift = $37.78 - 24.44 = 13.34$ k

Net uplift = gross uplift − pile self-weight = $13.34 - 5.6 = 7.74$ k          **D**

## 217

Factored load: $P_u = 1.2D + 1.6L = 1.2 \times 50 + 1.6 \times 90 = 204$ k

Under factored load, uniform soil pressure, $q_u = 204/49 = 4.16$ k/ft$^2$

Pressure due to soil and footing weight = $120 \times 3.5 = 420$ psf = $0.42$ k/ft$^2$

Total soil pressure = $4.16 + 0.42 = 4.58$ k/ft$^2$

For no shear reinforcement, allowable shear stress in punching shear = $4\phi\sqrt{f'_c} = 178$ psi

Try footing thickness = 18 in.; effective depth = $18 - 4 = 14$ in.

Punching shear critical section located $d/2 = 7$ in. from face of column, creating a perimeter 28 in. × 28 in.

Total area outside critical perimeter (contributing to 2-way shear) = $84^2 - 28^2 =$ 6,272 sq. in. = 43.56 sq. ft

Total shear force outside critical perimeter = $43.56 \times 4.58 = 199.5$ k

Shear stress on this critical plane: $v_u = \dfrac{199.5}{4 \times 28 \times 14} = 0.127$ ksi $= 127$ psi $< 178$ psi

Try footing thickness = 15 in.; effective depth = 15 − 4 = 11 in.

Punching shear critical section located $d/2 = 5.5$ in. from face of column, creating a perimeter 25 in. × 25 in.

Total area outside critical perimeter (contributing to 2-way shear) = $84^2 - 25^2 =$ 6,431 sq. in. = 44.66 sq. ft.

Total shear force outside critical perimeter = $44.66 \times 4.58 = 204.5$ k

Shear stress on this critical plane: $v_u = \dfrac{204.5}{4 \times 25 \times 11} = 0.186$ ksi $= 186$ psi $< 178$ psi

Smallest acceptable footing thickness = 18 in. **B**

## 218

Compared to the span (390 ft), the spacing between vertical hangers (6 ft) is small. Therefore, the loading on the cable may be considered uniformly distributed. Therefore, the deformed shape of the cable will be parabolic.

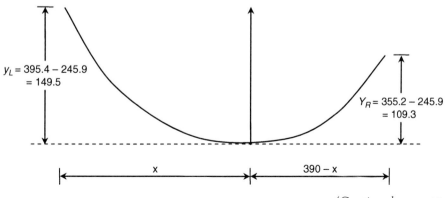

(*Continued on next page*)

With respect to the assumed origin at the low point (assume distance from tower 1 = x):

$$y_L = 395.4 - 245.9 = 149.5 \text{ ft} = c(-x)^2$$

$$y_R = 355.2 - 245.9 = 109.3 \text{ ft} = c(390-x)^2$$

$$\frac{390-x}{x} = \sqrt{\frac{109.3}{149.5}} = 0.855 \Rightarrow x = 210.24 \text{ ft}$$

UDL on cable: $w = 3.2$ k/ft (horizontal projection)

With respect to origin at the low point of the cable, the springing point of the cable (top of supporting tower 1) has coordinates: $x = -210.24$ ft, $y = 149.5$ ft. The equation of the cable is:

$$y = \frac{wx^2}{2H} \Rightarrow H = \frac{wx_o^2}{2y_o} = \frac{3.2 \times 210.24^2}{2 \times 149.5} = 473.1 \text{ k} = 236.5 \text{ tons}$$

The maximum tension at the top of tower 1 (furthest point from low point of cable) is:

$$T_{max} = \sqrt{H^2 + w^2 x^2} = \sqrt{473.1^2 + 3.2^2 \times 210.24^2} = 822.5 \text{ k} = 411.2 \text{ tons} \qquad \textbf{D}$$

## 219

Bending moment: $M_{max} = \dfrac{wL^2}{8} = \dfrac{275 \times 20^2}{8} = 13,750 \text{ lb} \cdot \text{ft} = 165,000 \text{ lb} \cdot \text{in.}$

For a sawn 6 × 10 timber beam, section modulus $S_x = 82.729$ in.$^3$ (dressed dimensions 5.5 × 9.5 in.)

Bending stress: $\sigma = \dfrac{M}{S} = \dfrac{165,000}{82.729} = 1,995 \text{ psi}$ \qquad **D**

## 220

The dressed dimensions of the column cross-section (2 × 6 nominal) are 1.5 in. × 5.5 in.

For buckling about the x-axis, $K = 2.10$ (recommended - NDS)

Slenderness ratio: $\dfrac{KL}{r} = \dfrac{2.1 \times 12 \times 12}{0.29 \times 5.5} = 189.6$

For buckling about the y-axis, $K = 0.80$ (recommended - NDS)

Slenderness ratio: $\dfrac{KL}{r} = \dfrac{0.8 \times 12 \times 12}{0.29 \times 1.5} = 264.8$

Euler buckling load, $P_E = \dfrac{\pi^2 EA}{\left(\dfrac{KL}{r}\right)^2} = \dfrac{\pi^2 \times 1.5 \times 10^6 \times 1.5 \times 5.5}{264.8^2} = 1{,}741 \text{ lb} = 1.74 \text{ k}$

The stress level is 211 psi (for most grades of lumber, this ensures elastic behavior).

Correct answer is A

## 221

The vertical reaction at A can be calculated by taking moments about G:

$$A_y = \frac{16 \times 75 + 26 \times 60 + 30 \times 45 + 26 \times 30 + 16 \times 15}{90} = 57 \text{ k}$$

However, note that in this case, the structure *and* the load are symmetric; therefore, the two reactions will be equal to half the total vertical load of 114 k. Also note that if the end-game (of taking moments about A) is seen ahead of time, then it is obvious that the vertical reaction at A will not be necessary and therefore one can avoid that calculation.

Taking moments about A (so that the intersecting forces $F_{CD}$ and $F_{KJ}$ are not involved):

$$\frac{4}{5}F_{CI} \times 30 + \frac{3}{5}F_{CI} \times 20 + 16 \times 15 + 26 \times 30 = 0 \Rightarrow F_{CI} = -28.3 \text{ k}$$

Correct answer is D

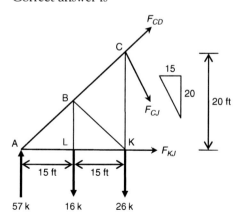

## 222

Maximum permitted reinforcement (ACI) = 8%

Factored load, $P_u = 1.2 \times 300 + 1.6 \times 350 = 920$ k

$$P_u \leq \phi\beta A_g [0.85f_c'(1-\rho_g)+\rho_g f_y]$$

$0.7\times0.85\times A_g \times [0.85\times4\times(1-0.08)+0.08\times60] = 4.717A_g \geq 920 \Rightarrow A_g \geq 195$ in.²

Diameter ≥ 15.8 in. Choose D = 16 in.      **C**

## 223

The HL93 loading consists of: lane load of 640 lb/ft per loaded lane (10-ft wide) + (HS20) truck load in a design lane.

Therefore, per design lane, the maximum shear on a simple span is given by:

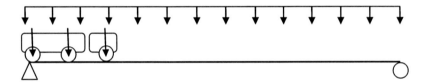

HL-93 loading = (HS25 truck or Tandem Load, whichever governs) + Lane Load

For the maximum shear at either support A or B, the distance between 2nd and 3rd wheels (which varies) should be minimum, that is, 14 ft.

Lane Load = 0.64 k/ft

Then, maximum shear = vertical reaction at the left support (for the load placement shown) = contribution of truck wheels (left to right) + contribution of lane load:

$$V_{max} = 32+\frac{125-14}{125}\times32+\frac{125-14-14}{125}\times8+\frac{0.64\times125}{2}=106.62 \text{ k} \qquad \mathbf{C}$$

For tandem (two 25-k axles separated by 4 ft), the maximum shear is given by:

$$V_{max} = 25+\frac{125-4}{125}\times25+\frac{0.64\times125}{2}=89.2 \text{ k}$$

## 224

For a simply supported beam with uniformly distributed load, the maximum (elastic) deflection is given by:

$$\delta_{max} = \frac{5wL^4}{384EI}$$

Moment of inertia of the dressed timber beam:

$$I = \frac{1}{12}bh^3 = \frac{1.5 \times 9.25^3}{12} = 98.93 \text{ in.}^4$$

Therefore,

$$\delta_{max} = \frac{5wL^4}{384EI} \leqq \frac{L}{240} \Rightarrow w \leqq \frac{384EI}{1,200L^3} = \frac{384 \times 1.5 \times 10^6 \times 98.93}{1,200 \times 168^3} = 10.01 \text{ lb/in.}$$
$$= 120.13 \text{ lb/ft}$$

Maximum spacing $S = 120.13$ plf/55 psf $= 2.18$ ft. $= 26.2$ in. Use joists at 24-in. spacing.　　　　　　　　　　　　　　　　　　　　　　　　　　　　　　　　　　　　C

## 225

Since the beam is braced at midspan, $L_{unbraced} = 30/2 = 15$ ft

*Note*: Use unbraced length to calculate the Maximum Factored Moment Capacity.

### LRFD

From the AISC Manual, for $L_b = 15$ ft, for the W16 × 100 section, $\phi_b M_n = 670$ k-ft.

Max. factored moment capacity $= 670$ k-ft (approx.)　　(Using AISC Manual Table 3-10)

For a simple span, maximum moment for uniformly distributed load:

$$M_u = \frac{w_u L^2}{8} \Rightarrow w_u = \frac{670 \times 8}{30^2} = 5.96 \text{ k/ft}$$

(*Continued on next page*)

## ASD

From the AISC manual, for $L_b = 15$ ft, for the W16 × 100 section, $M_n/\Omega = 450$ k-ft.

Max. moment capacity = 450 k-ft (approx.)     (Using AISC Manual Table 3-10)

For a simple span, maximum moment for uniformly distributed load:

$$M_a = \frac{wL^2}{8} \Rightarrow w = \frac{450 \times 8}{30^2} = 4.0 \text{ k/ft}$$

Therefore, the correct answer is                                                                                  **B**

## 226

Factored axial load: $P_u = 1.2P_D + 1.6P_L = 1.2 \times 300 + 1.6 \times 180 = 648$ k

Factored moment, $M_u = 1.2P_D e_D + 1.6P_L e_L = 1.2 \times 300 \times 0 + 1.6 \times 180 \times 4 = 1{,}152$ k · in.

The dimensionless parameters:

$$K_n = \frac{P_u}{\phi_c f_c' A_g} = \frac{648}{0.65 \times 4 \times 320} = 0.78$$

$$R_n = \frac{M_u}{\phi_c f_c' A_g h} = \frac{1{,}152}{0.65 \times 4 \times 320 \times 20} = 0.07$$

Assuming effective cover of about 2.5 in., $\gamma h = 20 - 2 \times 2.5 = 15$ in. Therefore, $\gamma = 0.75$.

For the above data $(f_c' = 4$ ksi, $f_y = 60$ ksi, $\gamma = 0.75)$, the column interaction diagram yields $\rho_g = 1.2\%$.

Therefore, $A_s = 0.012 \times 16 \times 20 = 3.84$ in.$^2$                                           **B**

## 227

The resultant horizontal load $= 0.2$ k/ft × 80 ft $= 16$ k is resisted by the two shear walls parallel to the y-axis, in proportion to their tributary area (simple beam model). Therefore, each wall carries a shear of 8 k.

The unit shear in wall number $2 = \dfrac{8{,}000}{45} = 177.8$ lb/ft                               **A**

## 228

### ASD Solution

Service loads are as shown on the figure. The vertical reaction at A can be calculated by taking moments about E:

$$A_y = \frac{120 \times 36 + 200 \times 24 + 120 \times 12 - 240 \times 9}{48} = 175 \text{ k}$$

Under these loads, $A_x = -240$ k, $A_y = 175$ k, $E_y = 265$ k

By beam analogy, the maximum tension in the bottom chord will occur at midspan (members BC and CD).

Using method of sections, $F_{BC} = 546.7$ k

Yield criterion: $0.6F_y A_g \geq 546.7 \Rightarrow A_g \geq \dfrac{546.7}{0.6 \times 36} = 25.3 \text{ in.}^2$

Fracture criterion: $0.5F_u A_e = 0.5F_u(0.75A_g) \geq 546.7 \Rightarrow A_g \geq \dfrac{546.7}{0.5 \times 0.75 \times 58} = 25.1 \text{ in.}^2$

### LRFD Solution

Factored loads are as follows:  at B and D: $1.2 \times 0.3 \times 120 + 1.6 \times 0.7 \times 120 = 177.6$ k
at C: $1.2 \times 0.3 \times 200 + 1.6 \times 0.7 \times 200 = 296$ k
at F: $1.2 \times 0.3 \times 240 + 1.6 \times 0.7 \times 240 = 355.2$ k

The vertical reaction at A can be calculated by taking moments about E:

$$A_y = \frac{177.6 \times 36 + 296 \times 24 + 177.6 \times 12 - 355.2 \times 9}{48} = 259 \text{ k}$$

Under these loads, $A_x = -355.2$ k, $A_y = 259$ k, $E_y = 392.2$ k

By beam analogy, the maximum tension in the bottom chord will occur at midspan (members BC and CD).

Using method of sections, $F_{BC} = 809.1$ k

Yield criterion: $0.9F_y A_g \geq 809.1 \Rightarrow A_g \geq \dfrac{809.1}{0.9 \times 36} = 24.97 \text{ in.}^2$

Fracture criterion: $0.75F_u A_e = 0.75F_u(0.75A_g) \geq 809.1 \Rightarrow A_g \geq \dfrac{809.1}{0.75^2 \times 58} = 24.8 \text{ in.}^2$

Correct answer is                                                                                                          **D**

## 229

Resultant wind force on sign: $R_{w1} = 8 \times 4 \times 28 = 896$ lb

Resultant wind pressure on cylindrical post: $R_{w2} = 0.7 \times 20 \times \dfrac{5.1}{12} \times 28 = 166.6$ lb

Total direct shear force at base of the post $= R_{w1} + R_{w2} = 1,062.6$ lb

Distance from center of hollow post to location of wind resultant $= 4$ ft $+ \dfrac{1}{2}(5.1)$ in. $=$ 50.55 in.

Torsional moment about axis of the post: $T = 896 \times 50.55 = 45,293$ lb $\cdot$ in.

Cross sectional area of post: $A = \dfrac{\pi}{4}(d_o^2 - d_i^2) = \dfrac{\pi}{4}(5.1^2 - 4.5^2) = 4.52$ in.$^2$

Polar moment of inertia of post: $J = \dfrac{\pi}{32}(d_o^4 - d_i^4) = \dfrac{\pi}{32}(5.1^4 - 4.5^4) = 26.16$ in.$^4$

For a hollow (thin-walled) circular tube, shear stress due to transverse shear is approximately:

$$\tau \approx \frac{2V}{A} = \frac{2 \times 1,062.6}{4.52} = 470 \text{ psi}$$

Shear stress due to torsion:

$$\tau = \frac{Tr}{J} = \frac{45,293 \times 2.55}{26.16} = 4,415 \text{ psi}$$

Maximum shear stress at base $= 470 + 4,415 = 4,885$ psi    **B**

## 230

For cantilever columns, which are not allowed to rotate at the "free" end, the load displacement relationship is $\Delta = PL^3/12EI$.

The stiffness is, therefore, $k = 12EI/L^3$.

The effective lumped weight is: $\frac{1}{2}W_{\text{column}} + W_{\text{floor}} = 0.5 \times 1.8 + 48 = 48.9$ k

Natural frequency:

$$\omega_n = \sqrt{\frac{k}{m}} = \sqrt{\frac{kg}{W}} = \sqrt{\frac{12EI}{L^3}\frac{g}{W}} = \sqrt{\frac{12 \times 29,000 \times 10,000 \times (32.2 \times 12)}{(14 \times 12)^3 \times 48.9}} = 76.2 \text{ rad/sec}$$

Fundamental period: $T_n = \dfrac{2\pi}{\omega_n} = 0.083$ sec

Correct answer is                                                                          A

## 231

For $f'_c$ greater than 5,000 psi, reduction of all content indicated in ACI 318 Manual Table 4.4.1 by 1.0%.

Based on ACI 318 Manual Table 4.4.1 (Air content) and Table 4.2.1 (Exposure condition), air content percentage = 6 − 1 = 5%                                                 D

## 232

The factored load is $w_u = 1.2\,w_{\text{DL}} + 1.6\,w_{\text{LL}} = 8.8$ k/ft

The design moment is $M_u = w_u L^2/8 = 687.5$ k-ft

We are looking for the *smallest* size of beam, therefore for *maximum allowed reinforcement*. The strength reduction factor is being assumed to be 0.81 (maximum steel → minimum $\phi$).

ACI318-05 gives $\phi = 0.48 + 83\varepsilon_t \le 0.9$. Minimum permissible tensile strain $\varepsilon_t = 0.004 \rightarrow \phi = 0.812$.

Using the design table (Table 105.5) in *All-in-One*:

For $f'_c = 4$ ksi, $f_y = 60$ ksi, $\rho_{\text{max}} = 0.0206$, $w_{\text{max}} = \dfrac{0.0206 \times 60}{4} = 0.309 \rightarrow X = 0.2527$

Thus, the effective depth: $d = \sqrt{\dfrac{M_u}{0.81Xbf'_c}} = \sqrt{\dfrac{687.5 \times 12}{0.81 \times 0.2527 \times 15 \times 4}} = 25.9$ in.

Assuming $d \approx h - 2.5$, minimum overall depth = 25.9 + 2.5 = 28.4 in.                  C

## 233

### Flange ratio

$$\frac{b_f}{2t_f} = \frac{12}{2 \times 2} = 3$$

Compact limit, $\lambda_p = 0.38\sqrt{\dfrac{E}{F_y}} = 0.38 \times \sqrt{\dfrac{29,000}{50}} = 9.15$    (AISC Table B4.1)

Since $\dfrac{b}{t} = 3 < \lambda_p$, therefore, the flange is compact.

### Web

$$\frac{h}{t_w} = \frac{48}{0.5} = 96$$

Compact limit: $\lambda_p = 3.76\sqrt{\dfrac{E}{F_y}} = 3.76 \times \sqrt{\dfrac{29,000}{50}} = 90.31$

Non-compact limit: $\lambda_r = 5.70\sqrt{\dfrac{E}{F_y}} = 5.70 \times \sqrt{\dfrac{29,000}{50}} = 137.27$

Since $\lambda_p < \dfrac{h}{t_w} < \lambda_r$, web is non-compact.

Since the web is non-compact, the whole section is non-compact.                    C

## 234

Since gravity forces counteract uplift, we must consider the empty tank case here. The weight of 6 k is carried equally by each leg. Therefore, the compression at each footing = 1.5 k.

The lateral force creates an overturning moment = 120 k × 65 ft = 7,800 k-ft on the horizontal plane at the top of the footings. This overturning moment is shared equally by two couples (4 legs in two pairs). Therefore, each resisting couple = 3,900 k-ft.

As the lever arm on each couple is 30 ft, the force at each footing = 3,900/30 = 130 k. This is an added compression under the legs on the far side (right) of the tower and an uplift under the legs on the near side (left).

Therefore, the near side legs experience a "net" uplift of 130 − 1.5 = 128.5 k          B

## 235

Taking moments about D:

$$\sum M_D = 8 \times 20 - 18B_y + 36 \times 4.5 - 12 \times 6 = 0$$

$$B_y = \frac{8 \times 20 + 36 \times 4.5 - 12 \times 6}{18} = 13.89$$

Using equilibrium of vertical forces, $D_y = 8 + 36 + 12 - 13.89 = 42.11$

The free body diagram is followed by the shear force and bending moment diagrams. The areas under the shear diagram are used to generate the bending moment diagram. For each segment, the order (constant, linear, quadratic, etc.) of the moment diagram is *one higher* than that of the shear diagram.

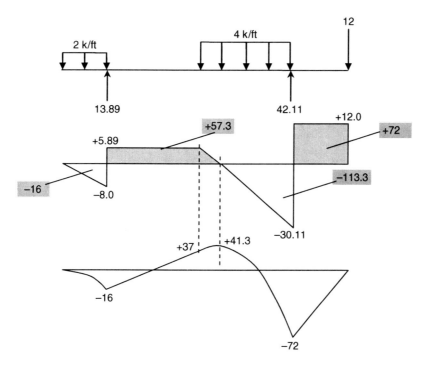

The maximum bending moment is 72 k-ft

B

## 236

According to ASCE 7-08, the correct answer choice is      A

## 237

Solve for truss member forces using equilibrium of joint B:

$$\sum F_x = 0 \Rightarrow F_{AB} = F_{CB}$$

$$\sum F_y = 0 \Rightarrow 2 \times \frac{10}{\sqrt{164}} \times F_{AB} = 30 \text{ k} \Rightarrow F_{AB} = +19.21 \text{ k (tension)}$$

$$F_{AB} = F_{BC} = +19.21 \text{ k (T)}$$

Create the virtual load (consistent with desired deflection).

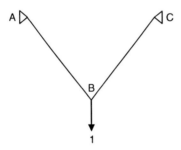

Calculate the member forces due to virtual load

*Note*: In this case, the real load and the Virtual load are congruent, so we can use scaling.

$$f_{AB} = f_{BC} = +19.21 \text{ k}/30 \text{ k} = +0.64 \text{ (T)}$$

According to the Principle of Virtual Work, the vertical deflection at B is given by:

$$\Delta = \sum \frac{FfL}{AE} = \frac{19.21 \times 0.64 \times (12.81 \times 12)}{2 \times 29,000} + \frac{19.21 \times 0.64 \times (12.81 \times 12)}{3 \times 29,000} = 0.054 \text{ in.}$$

Correct answer is      A

*Note*: The positive sign of the answer indicates that this deflection is in the same sense as the assumed unit load (downward).

## 238

Knowledge of the nodal moments $M_{AB}$, $M_{BA}$, $M_{BC}$, etc. allows the "decoupling" of the individual spans, as shown below.

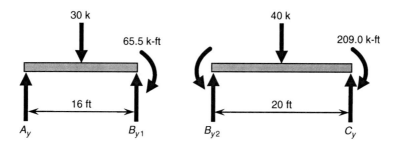

For the left substructure, taking moments about A yields $B_{y1}$:

$$M_A = 30 \times 8 + 65.5 - 16B_{y1} = 0 \Rightarrow B_{y1} = 19.094 \text{ k}$$

For the right substructure, taking moments about C yields $B_{y2}$:

$$M_C = 209 - 65.5 - 40 \times 10 + 65.5 + 20B_{y2} = 0 \Rightarrow B_{y2} = 12.825 \text{ k}$$

Therefore, total vertical reaction at $B = B_{y1} + B_{y2} = 19.094 + 12.825 = 31.919 \text{ k}$     D

## 239

The effective width of flange = smallest of (L/4, c.c spacing, $b_w + 12t$) = min (7 ft, 8 ft, 88 in.) = 7 ft = 84 in.

Cross section of beam & slab = $96 \times 5 + 28 \times 15 = 900 \text{ in.}^2 = 6.25 \text{ ft}^2$

Self weight of beam & slab = $6.25 \text{ ft}^2 \times 0.15 \text{ k/ft}^3 = 0.94 \text{ k/ft}$

Floor live load transmitted to each beam = 85 psf × 8 ft = 680 lb/ft = 0.68 k/ft

Total factored load on beam = $1.2 \times 0.94 + 1.6 \times 0.68 = 2.216 \text{ k/ft}$

Maximum moment: $M_u = \dfrac{wL^2}{8} = \dfrac{2.216 \times 28^2}{8} = 217.17 \text{ k} \cdot \text{ft} = 2{,}606 \text{ k} \cdot \text{in.}$

*(Continued on next page)*

For a rectangular beam with ($b = 84$ in., effective depth $d = 17.5$ in., $M_u = 2{,}606$ k-in., $f_c' = 4$ ksi, and $f_y = 60$ ksi):

Strength parameter: $X = \dfrac{M_u}{\phi f_c' b d^2} = \dfrac{2{,}606}{0.9 \times 4 \times 84 \times 17.5^2} = 0.0281$

From the table (Table 105.5 in *All-in-One*), corresponding value of the reinforcement parameter:

$$w = \rho \frac{f_y}{f_c'} = 0.0285$$

$$\rho = w \frac{f_c'}{f_y} = \frac{0.0285 \times 4}{60} = 0.0019$$

This is less than the minimum steel ratio, so according to ACI, provide 33% more than required steel.

Area of steel, $A_s = 1.33 \times 0.0019 \times 84 \times 17.5 = 3.72$ in.$^2$      **B**

## 240

The correct answer choice is      **C**

# Answer Key for Structural Deapth Exam

| | | | | | | | | | |
|---|---|---|---|---|---|---|---|---|---|
| 201 | D | | 211 | A | | 221 | D | | 231 | D |
| 202 | C | | 212 | C | | 222 | C | | 232 | C |
| 203 | B | | 213 | C | | 223 | C | | 233 | C |
| 204 | C | | 214 | B | | 224 | C | | 234 | B |
| 205 | D | | 215 | B | | 225 | B | | 235 | B |
| 206 | C | | 216 | D | | 226 | B | | 236 | A |
| 207 | A | | 217 | B | | 227 | A | | 237 | A |
| 208 | A | | 218 | D | | 228 | D | | 238 | D |
| 209 | D | | 219 | D | | 229 | B | | 239 | B |
| 210 | C | | 220 | A | | 230 | A | | 240 | C |

# 11

## Geotechnical Depth Exam Solutions

These detailed solutions are for questions 301 to 340, representative of a 4-hour geotechnical depth exam according to the syllabus and guidelines for the Principles and Practice (P&P) of Civil Engineering Examination, administered by the National Council of Examiners for Engineering and Surveying (NCEES).

### 301

By examination, it seems that the peak dry unit weight will come from sample 3 or 4.

Volume of Standard Proctor mold = 1/30 ft$^3$

For sample 3, total unit weight, $\gamma = W/V = 3.95/(1/30) = 118.5$ lb/ft$^3$

Dry unit weight, $\gamma_d = 118.5/1.16 = 102.2$ lb/ft$^3$

For sample 4, total unit weight, $\gamma = W/V = 4.21/(1/30) = 126.3$ lb/ft$^3$

Dry unit weight, $\gamma_d = 126.3/1.18 = 107.0$ lb/ft$^3$ (maximum dry unit weight)

Target dry unit weight = $0.9 \times 107 = 96.3$ lb/ft$^3$

(Continued on next page)

Weight of solids needed $= 96.3 \times 1.5 \times 10^6 \times 27 = 3.9 \times 10^9$ lb

From table, for 12% moisture content, unit weight $= 3.24/(1/30) = 97.2$ lb/ft³;
$\gamma_d = 97.2/1.12 = 86.8$ lb/ft³

Volume of borrow soil needed $= 3.9 \times 10^9$ lb/86.8 lb/ft³ $= 4.49 \times 10^7$ ft³ $= 1.66 \times 10^6$ yd³

B

# 302

### Sample 1 (undrained)–effective stress analysis

Total axial stress at failure is calculated as a sum of the radial chamber pressure and the added load of 158.2 lb.

$$\sigma_1 = 18 + \frac{158 \cdot 2}{\frac{\pi}{4}(2)^2} = 68.36 \text{ psi}$$

Effective vertical stress is given by:

$$\sigma_1' = 68.36 - 5.6 = 62.76 \text{ psi}$$

Effective radial stress is given by:

$$\sigma_3' = 18 - 5.6 = 12.4 \text{ psi}$$

For clay, assuming angle of internal friction $= 0$, cohesion:

$$c = \frac{\sigma_1' - \sigma_3'}{2} = \frac{62.76 - 12.4}{2} = 25.18 \text{ psi}$$

### Sample 2 (drained, pore pressure zero)–total stress analysis

Radial stress, $\sigma_3' = 36$ psi

$$c = \frac{\sigma_1' - \sigma_3'}{2} \Rightarrow \sigma_1' = \sigma_3' + 2c = 36 + 2 \times 25.18 = 86.36 \text{ psi}$$

As there is no pore pressure, this is also the value of the total vertical stress. Therefore, the axial deviatoric stress $= 86.36 - 36 = 50.36$ psi.

Corresponding axial load at failure: $P_f = 50.36 \times \frac{\pi}{4} \times 2^2 = 158.2$ lb

D

## 303

Eccentricity: $e = M/P = 9/10 = 0.9$ ft

In order to determine whether the entire footing width is effective, the eccentricity must be compared to $B/6$, which is $4/6 = 0.67$ ft.

Given that this eccentricity is greater than $B/6$, the resultant load does not fall within the middle third of the footing base. This will cause uplift on the far side of the footing. Maximum soil pressure is given by:

$$q_{max} = \frac{4P}{3(B-2e)} = \frac{4 \times 10}{3 \times (4-2 \times 0.90)} = 6.06 \text{ ksf}$$

**C**

## 304

All lateral and vertical force components are shown on the figure. A vertical plane is drawn through the heel of the footing.

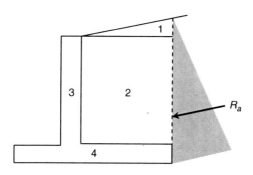

The height H, measured vertically from the bottom of the footing to the top of the backfill, $H = 17 + 6 \tan 15 = 18.6$ ft.

For $\phi = 34°$, $\beta = 15°$, $\delta = 0°$, $\theta = 90°$.

Rankine's active earth pressure coefficient, $K_a = 0.339$

Earth pressure resultant: $R_a = \frac{1}{2}K_a \gamma H^2 = 0.5 \times 0.339 \times 120 \times 18.6^2 = 7,037$ lb/ft.

*(Continued on next page)*

This can be resolved into its vertical and horizontal components as below:

$R_{av} = 7,037 \times \sin 15 = 1,821$ lb/ft

$R_{ah} = 7,037 \times \cos 15 = 6,797$ lb/ft

The weights of the soil wedges 1, 2 and the concrete elements 3 and 4 are calculated below, with their horizontal distances (measured from the toe).

$W_1 = \dfrac{1}{2} \times 6 \times 1.6 \times 120 = 576$ @ 9 ft

$W_2 = 14 \times 6 \times 120 = 10,080$ @ 8 ft

$W_3 = 1 \times 14 \times 150 = 2,100$ @ 4.5 ft

$W_4 = 11 \times 3 \times 150 = 4,950$ @ 5.5 ft

$R_{av} = 1,821$ @ 11 ft

The sum of the stabilizing moments about the toe is given by:

$M_s = \sum F_i x_i = 576 \times 9 + 10,080 \times 8 + 2,100 \times 4.5 + 4,950 \times 5.5 + 1,821 \times 11$

$$= 142,530 \text{ lb} \cdot \text{ft/ft}$$

The overturning moment about the toe is given by:

$M_{OT} = R_{ah} \times H/3 = 6,797 \times 18.6/3 = 42,141$ lb·ft/ft

$FS_{OT} = 142,530/42,141 = 3.38$           C

## 305

At a depth of 30 ft, effective vertical stress is given by:

$\sigma'_v = 124 \times 10 + (124 - 62 \cdot 4) \times 20 = 2,472$ psf

Average cyclic shear stress: $\tau_{ave} = CSR \times \sigma'_v = 0.23 \times 2,472 = 568.6$ psf

Ultimate shear strength of the soil in the field (corrected for relative density) is:

$\tau_{ult} = 0.85/0.95 \times 1,200 = 1,074$ psf

Factor of safety for liquefaction is given by:

$$FS = \frac{\tau_{ult}}{\tau_{ave}} = \frac{1{,}074}{568.6} = 1.89$$

B

## 306

The dead load + wind load case produces maximum uplift.

Total dead load = 660 k, which produces compression in each pile = 660/27 = 24.4 k.

Wind load overturning moment is calculated as the sum of the moments of lateral forces about the top of the pile cap.

$$M_{OT} = 30 \times 15 + 60 \times 30 + 70 \times 45 + 80 \times 60 = 10{,}200 \text{ k} \cdot \text{ft}$$

Assuming linear force distribution in the columns (spacing $S$) and taking moments about pile line 5, we have 6 piles (lines 4 and 6) at distance $S$, 6 piles (lines 3 and 7) at distance $2S$, 6 piles (lines 2 and 8) at distance $3S$, and 6 piles (lines 1 and 9) at distance $4S$. The sum of these moments = $6 \times F \times S + 6 \times 2F \times 2S + 6 \times 3F \times 3S + 6 \times 4F \times 4S = 180FS = 10{,}200$.

$F = 10{,}200/(180 \times 6) = 9.44$ k. Thus, each pile in the outermost line of piles experiences a wind-uplift = $4F = 37.78$ k.

Gross uplift = 37.78 − 24.44 = 13.34 k

Net uplift = gross uplift − pile self-weight = 13.34 − 5.6 = 7.74 k

A

## 307

Effective stress at the center of the clay layer (depth = 14 ft) is calculated as the sum of the $\Sigma\gamma'z$ products for the soil layers (submerged where appropriate).

$$p_1' = 114 \times 6 + (120 - 62.4) \times 2 + (108 - 62.4) \times 6 = 1{,}073 \text{ psf}$$

The uniform pressure directly under the mat (depth = 3.5 ft) is:

$$\Delta p = \frac{20{,}000 \times 2{,}000}{190 \times 250} = 842 \text{ psf}$$

(Continued on next page)

The pressure increase at depth of 14 ft (10.5 ft below bottom of mat) can be considered to be 842 psf (no dissipation, since the load is exerted over relatively large plan dimensions). The new pressure resulting from stress relief due to excavation for the mat foundation, lowering of water table and application of the foundation load is:

$$p_2' = 114 \times 8 + (108 - 62.4) \times 6 - 399 + 842 = 1,629 \text{ psf}$$

Primary settlement of the 12-ft-thick clay layer is calculated as:

$$s = \frac{C_c H \log_{10}\left(\frac{p_2'}{p_1'}\right)}{1 + e_0} = \frac{0.45 \times 144 \text{ in.}}{1 + 1.4} \times \log_{10}\left(\frac{1,629}{1,073}\right) = 4.9 \text{ in.} \qquad \text{A}$$

## 308

The clay layer can be considered to be "singly drained" because the underlying clay layer can be considered impermeable. Therefore, the drainage thickness $H_d$ = the full layer thickness = 12 ft.

For degree of consolidation $U = 80\%$, time factor $T_v = 0.567$.

The time for 80% settlement to occur is given by:

$$t = \frac{T_v H_d^2}{c_v} = \frac{0.567 \times 12^2}{6} = 13.6 \text{ years} \qquad \text{C}$$

## 309

Weight of wet soil = 1,331.5 g

As the water content = 15.2%, this can be split into dry soil solids = 100/115.2 × 1,331.5 = 1,155.8 g and water (175.7 g).

Since the SG of soil solids is known (2.70), we also have volume of soil solids = 1,155.8/2.7 = 428.07 cm³ and volume of water = 175.7 cm³.

Also, the buoyancy (apparent loss of weight on immersion) of the wax-coated sample = 1,368.2 − 593.4 = 774.8 g. Therefore, the volume of displaced water = volume of wax-coated sample = 774.8 cm³. (Density of water = 1 g/cm³)

The weight of the wax = 1,368.2 − 1,331.5 = 36.7 g. Using the specific gravity of the wax, we can calculate the wax volume = 36.7/0.9 = 40.78 cm³.

Therefore, the volume of the (uncoated) soil sample = 774.8 − 40.78 = 734.02 cm³

Therefore, volume of air = 734.02 − (428.07 + 175.7) = 130.25 cm³

Total volume of voids in the original sample = 175.7 + 130.25 = 305.95 cm³

Degree of saturation, $S = V_{water}/V_{voids}$ = 175.7/305.95 = 0.574          A

## 310

Weight of test sand in test hole = 13.75 − 10.24 = 3.51 lb

Volume of test sand in test hole = 3.51/88.2 = 0.0398 ft³

Weight of soil obtained from the test hole = 5.52 lb

Unit weight of soil obtained from the test hole = 5.52/0.0398 = 138.71 lb/ft³

Dry unit weight of soil = 138.71/1.19 = 116.56 lb/ft³

Percent compaction = 116.56/126.3 = 92%          A

## 311

The hydraulic gradient in the sand drain is calculated from the head difference and the total length:

$$i = \frac{H}{L} = \frac{945 - 905}{160} = 0.25 \text{ ft/ft}$$

Sand porosity: $n = \dfrac{e}{1+e} = \dfrac{0.45}{1+0.45} = 0.31$

Seepage velocity (this is the true fluid velocity through the voids in the soil, rather than an effective velocity across the entire cross section) is given by:

$$v_s = \frac{Ki}{n} = \frac{1 \times 10^{-4} \times 0.25}{0.31} = 8.1 \times 10^{-5} \text{ ft/sec}$$

Scour velocity = 8 in./hr = $1.85 \times 10^{-4}$ ft/sec

Factor of safety against scour is given by:

$$FS = \frac{1.85 \times 10^{-4}}{8.1 \times 10^{-5}} = 2.29$$          C

## 312

The head difference (top of first sand layer to bottom of clay layer) = 18 in.

Hydraulic gradient = 18/12 = 1.5

Porosity: $n = \dfrac{e}{1+e} = \dfrac{0.7}{1.7} = 0.412$

Seepage velocity: $V_s = \dfrac{Ki}{n} = \dfrac{1.2 \times 10^{-5} \times 1.5}{0.412} = 4.369 \times 10^{-5}$ ft/hr = 0.0126 in./day

Time for leachate to seep through the 12-in. layer = 12/0.0126 = 954 days            **A**

*Note:* If the hydraulic conductivity of clay is $K$ and that of sand is $100K$, then the head loss in the clay layer is 99% of the total head loss. The head loss experienced in the sand layers is negligible.

## 313

The cumulative % retained on the no. 200 sieve is 74.

$F_{200} = 100 - R_{200} = 100 - 74 = 26$

Since $F_{200} < 50$, soil is predominantly coarse. Therefore, first letter is S or G.

Coarse fraction = $100 - F_{200} = 74$. Therefore, half the coarse fraction is 37.

$R_4 = 8$ is less than half the coarse fraction. Therefore, first letter is S.

Since $F_{200} > 12$, second letter of classification should be based on plasticity characteristics rather than gradation. Use the Casagrande Plasticity chart with LL = 43, PI = 43 − 21 = 22. This lies above the A-line (more like clay than silt).

Soil is classified as SC.            **D**

## 314

Directly under the footing, the vertical pressure $p_o$ = 100 k/25 ft$^2$ = 4 ksf = 4,000 psf

Since the diagonal of the square footing is 7.07 ft, the radial distance of a corner from the center of the footing = 7.07/2 = 3.54 ft

Solution using Boussinesq's equation (applicable for $z \geq 2B$): $z = 10$ ft, $r = 3.54$ ft

$$q = \frac{3P}{2\pi z^2 \left[1 + \left(\frac{r}{z}\right)^2\right]^{5/2}} = \frac{3 \times 100}{2\pi \times 10^2 \left[1 + \left(\frac{3.54}{10}\right)^2\right]^{5/2}} = 0.355 \ \text{k/ft}^2$$

Solution using stress contour charts based on Boussinesq's equation:

At a depth of $z = 10$ ft under the corner of the $B = 5$-ft footing, look for the stress contour at depth $z = 10$ ft $= 2B$ and $r = 3.54$ ft $= 0.71$ B.

Point plots close to the 0.08p contour.

Pressure $= 0.08 \times 4,000 = 320$ lb/ft$^2$ 

B

# 315

Dimensions of pile group (between outside edges) = 22 ft × 10 ft

It is presumed that the pile group load transfers to the soil at approximately 2/3 of the pile length.

2/3 L of piles is at depth = 30.67 ft

Bottom of clay layer at depth = 54 ft

Therefore, thickness of clay deposit undergoing consolidation settlement = 54 – 30.67 = 23.33 ft. Stress propagation through half this thickness = 11.67 ft.

Using a 2:1 stress propagation pyramid, where a loaded area B × L gets distributed over an area (B + z)(L + z), distribute pile group load over an area (22 + 11.67) (10 + 11.67) = 33.67 ft × 21.67 ft.

$$\Delta p = \frac{125 \times 2,000}{33.67 \times 21.67} = 342.6 \ \text{psf}$$

Initial effective vertical stress: $p'_1 = 120 \times 20 + 115 \times 17 = 4,355 \ \text{lb/ft}^2$

After construction, effective vertical stress: $p'_2 = 4,355 + 342.6 = 4,697.6 \ \text{lb/ft}^2$

*(Continued on next page)*

Consolidation settlement:

$$s = \frac{HC_c}{1+e_0} \log_{10}\left(\frac{p_2'}{p_1'}\right) = \frac{23.33 \times 12 \times 0.4}{1.44} \log_{10}\left(\frac{4,697.6}{4,355}\right) = 2.56 \text{ in.}$$

B

## 316

Using the Taylor Stability Chart (shown below), using $\beta = 30$, $D = 50/20 = 2.5$, Stability number, $N_0 = 5.6$.

Factor of safety (slope stability): $FS = \dfrac{N_0 c}{\gamma H} = \dfrac{5.6 \times 900}{120 \times 20} = 2.1$

C

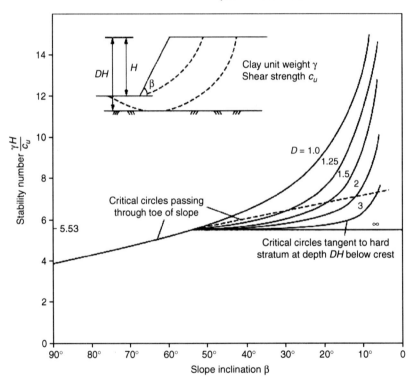

## 317

The compacted soil in the embankment has dry density $0.95 \times 95 \text{ pcf} = 90.25 \text{ pcf}$

500,000 yd³ of embankment volume contains solids $= 500,000 \times 27 \times 90.25 = 1.218 \times 10^9 \text{ lb}$

The borrow soil (while being transported) has a total density of 115 pcf and water content of 20%. Therefore, the dry density during transport = 115/1.2 = 95.83 pcf.

Total weight of soil during transport = $1.20 \times 1.218 \times 10^9 = 1.462 \times 10^9$ lb

Therefore, the volume of the soil (in transport) = $1.218 \times 10^9$ lb/95.83 = $1.271 \times 10^7$ ft$^3$

Truck carries V = 10 yd$^3$ = 270 ft$^3$, which weighs $270 \times 115 = 31,050$ lb

Number of trips based on weight = $1.462 \times 10^9$ lb/31,050 lb = 47,072

Number of trips based on volume = $1.271 \times 10^7$ ft$^3$/270 ft$^3$ = 47,072      **B**

## 318

$$K_a = \frac{1 - \sin\phi}{1 + \sin\phi} = 0.307$$

At depth of 6 ft, effective vertical pressure = $123 \times 6 = 738$ psf

Effective horizontal pressure = $0.307 \times 738 = 226.6$ psf

Total horizontal pressure = $226.6 + 0 = 226.6$ psf

At depth of 35 ft, effective vertical pressure = $123 \times 6 + (123 - 62.4) \times 29 = 2,495.4$ psf

Effective horizontal pressure = $0.307 \times 2,495.4 = 766.1$ psf

Total horizontal pressure = $766.1 + 62.4 \times 29 = 2,575.7$ psf

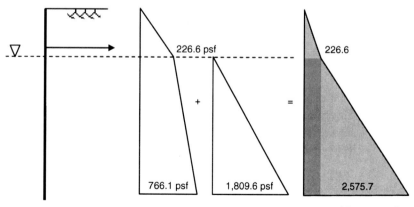

(*Continued on next page*)

Therefore, the total horizontal earth pressure diagram grows from zero at the surface, to 226.6 psf at a depth of 6 ft, to 2,575.7 psf at a depth of 35 ft. This can be broken up into a triangle from $z = 0$ to $z = 6$, a rectangle from $z = 6$ to $z = 35$, and a triangle from $z = 6$ to $z = 35$. The resultants of these pressure diagrams are calculated (in order) below:

Resultant earth pressure $= \dfrac{1}{2} \times 226.6 \times 6 + 226.6 \times 29 + \dfrac{1}{2} \times 29 \times 2,349.1 =$ 41,313 lb/ft

**C**

## 319

For clays, the cohesion is half the unconfined compression strength: $c = S_{uc}/2 = 600$ psf

Stiffness parameter: $\gamma H/c = 115 \times 24/600 = 4.6 > 4.0$. Therefore, assume it to be soft clay. According to Peck, the lateral pressure in a braced cut in soft clay grows linearly for the upper $H/4$ of the trench depth and then stays constant at that maximum value, which is approximately given by $\gamma H - 4c = 115 \times 24 - 4 \times 600 = 360$ psf.

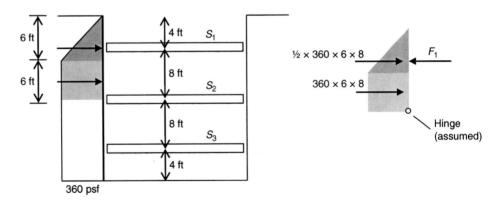

Taking moments about $S_2$ (assume hinge at second strut location), the moment of the earth pressure resultants (shown shaded) about $S_2$ is balanced by the moment of the strut compression. Note that the earth pressure terms include the out-of-plane 8-ft strut spacing

$$360 \times 6 \times 8 \times 3 + \dfrac{1}{2} \times 360 \times 6 \times 8 \times 8 = F_1 \times 8 \Rightarrow F_1 = 15{,}120 \text{ lb}$$

**B**

## 320

This solution will utilize the Theory of Buried Rigid Pipes (Marston and Spangler).

Aspect ratio of cover soil: $H/B_d = 14/8 = 1.75$

For this aspect ratio, the coefficient $C_d = 1.4$ (for sand and gravel)

Load: $q = C_d \gamma B_d^2 = 1.4 \times 125 \times 8^2 = 11{,}200$ lb/ft

D

## 321

Moment of inertia of pile group about group centerline (calculated as the second moment of all areas, $I = \Sigma A.d^2$)

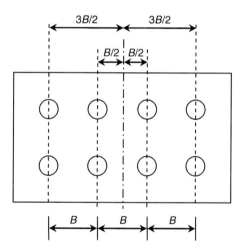

$$I = 4 \times A_p \times \left(\frac{B}{2}\right)^2 + 4 \times A_p \times \left(\frac{3B}{2}\right)^2 = 10 A_p B^2$$

The maximum compressive force in outer piles (pair to the right):

$$F = A\sigma = A\left(\frac{P}{nA} + \frac{M \times 1.5B}{10AB^2}\right) = \frac{P}{n} + \frac{M \times 1.5}{10B} = \frac{350}{8} + \frac{0.15 \times 300}{B}$$

Equating this to 30 tons = 60 k, we get $B_{min} = 2.77$ ft. Therefore, $B = 3$ ft 0 in.　　**B**

## 322

Head difference: $H = 213.45 - 198.65 = 14.80$ ft

Number of flow channels = 3; number of potential drops = 8

By Darcy's law, seepage flow per unit width is given by:

$$q = K\frac{N_f}{N_e}H = 200\times\frac{3}{8}\times14.8 = 1,110\ \frac{ft^2}{day} = 0.01285\ \frac{ft^2}{sec}$$

Total seepage loss under dam: $Q = qL = 0.01285\times150 = 1.93\ \dfrac{ft^3}{sec}$

C

## 323

With the porous end plates, pore pressures do not build up. Effective stress is equal to total stress. Completing the table with the normal ($\sigma = N/A$) and shear ($\tau = F/A$) stresses, we have:

| Sample | N (lb) | F (lb) | $\sigma$ (psi) | $\tau$ (psi) |
|--------|--------|--------|----------------|--------------|
| 1 | 120 | 78 | 9.55 | 6.21 |
| 2 | 160 | 93 | 12.73 | 7.40 |
| 3 | 220 | 116 | 17.51 | 9.23 |

Using sample points 1 & 3, the cohesion is calculated as:

$$c = \frac{\sigma_2\tau_1 - \sigma_1\tau_2}{\sigma_2 - \sigma_1} = \frac{17.51\times6.21 - 9.55\times9.23}{17.51 - 9.55} = 2.587\ psi = 372.5\ psf$$

A

## 324

Weight of the concrete wall: $W = (13\times1+8\times1)\times150 = 3,150$ lb/ft

Weight of the soil block above the heel: $W = (13\times3.5)\times118 = 5,369$ lb/ft

Total weight = 8,519 lb/ft

Maximum friction force mobilized = $0.5 \times 8,519 = 4,260$ lb/ft

$K_A = 0.3$

Active earth pressure resultant: $R_a = \frac{1}{2}K_a\gamma H^2 = \frac{1}{2}\times0.3\times118\times14^2 = 3{,}469$ lb/ft

Factor of safety against sliding instability, FS = 4,260/3,469 = 1.23     **D**

## 325

Since the question asks for ultimate load, and since concrete design is performed according to Strength Design Method in ACI 318, the factored column load is:

$w_u = 1.2D + 1.6L = 1.2\times125 + 1.6\times175 = 430$ k

Under factored load, uniform soil pressure: $q_u = \dfrac{430}{100} = 4.3$ ksf

Pressure due to soil and footing weight = $100 \times 5 = 500$ psf = 0.5 ksf

Ultimate 2-way punching shear = 4.8 ksf $\times$ 92.5 ft$^2$ = 444 k     **C**

## 326

Initial effective vertical stress at point P:

$p_1' = 114\times6 + (120-62.4)\times2 + (108-62.4)\times6 = 1{,}072.8$ psf

After building load, effective vertical stress at point P: $p_2' = 1{,}072.8 + 670 = 1{,}742.8$ psf

Settlement: $s = \dfrac{H\Delta e}{1+e_o} = \dfrac{HC_c}{1+e_o}\log_{10}\dfrac{p_2'}{p_1'} = \dfrac{144\times0.6}{1+1.4}\log_{10}\dfrac{1{,}742.8}{1072.8} = 7.6$ in.     **A**

## 327

Depth of water table below footing is 5 ft, which is less than the footing width (10 ft). Therefore, it *will* reduce the bearing capacity.

Since the water table is below the bottom of the footing, the average unit weight to be used in the third term of the bearing capacity equation is given by:

$\gamma_{ave} = \dfrac{\gamma D + (\gamma - \gamma_w)(B-D)}{B} = \dfrac{115\times5 + (115-62.4)\times(10-5)}{10} = 83.8$ pcf

(*Continued on next page*)

Ultimate bearing capacity (assuming given bearing capacity factors include shape factors) is:

$$q_{ult} = cN_c + \gamma D_f N_q + \frac{1}{2}\gamma_{ave}BN_\gamma = 115 \times 2 \times 25 + 0.5 \times 83.8 \times 10 \times 19 = 1,3711 \text{ psf}$$

Soil pressure due to column load + soil overburden = $750,000/100 + 115 \times 2 =$ 7,730 psf

FS = 13,711/7,730 = 1.77                                          B

## 328

Design capacity = 40 tons

Since FS = 6, ultimate capacity = $40 \times 6 = 240$ tons = 480,000 lb

Pile hammer energy WH = 50,000 ft-lb = 600,000 in.-lb

$$Q_{ult} = \frac{WH}{S+1.0} \Rightarrow S = \frac{WH}{Q_{ult}} - 1 = \frac{600,000}{480,000} - 1 = 0.25$$

$$S = 0.25 \frac{in.}{blow} \Rightarrow 4 \frac{blows}{in.} = 48 \frac{blows}{ft}$$                D

## 329

The theory for the falling head test can still be used as long as the area of the tube in which the liquid column is falling ($a$) is measured normal to the direction of head measurement (vertical). This formula is shown below ($h_1$ and $h_2$ are initial and final hydraulic head):

$$\ln\left(\frac{h_1}{h_2}\right) = \frac{KA(t_2 - t_1)}{aL}$$

The diameter of the horizontal section through the tube: $d' = \dfrac{d}{\sin 60}$

$$a = \frac{\pi d'^2}{4} = \frac{\pi d^2}{4\sin^2 60} = \frac{\pi 0.2^2}{4 \times 0.75} = 0.0419 \text{ in.}^2$$

$$D = \frac{\pi D^2}{4} = \frac{\pi(30^2)}{4} = 706.9 \text{ in.}^2$$

35 min = 0.583 hr

$$K = \frac{aL\ln\left(\dfrac{h_1}{h_2}\right)}{A(t_2 - t_1)} = \frac{0.0419 \times 20 \times \ln\left(\dfrac{35}{25}\right)}{706.9 \times 0.583} = 0.000684 \text{ in./hr} = 5.7 \times 10^{-5} \text{ ft/hr}$$

C

## 330

Active earth pressure coefficient ($\beta = 0$, $\theta = 90$, $\delta = 0$, $\phi = 34$): $K_a = 0.283$

Active earth pressure resultant: $R_a = \dfrac{1}{2}K_a\gamma H^2 = 0.5 \times 0.283 \times 125 \times 15^2 = 3{,}980$ lb/ft

Overturning moment: $M_a = \dfrac{1}{6}K_a\gamma H^3 = \dfrac{1}{6} \times 0.283 \times 125 \times 15^3 = 19{,}898.4$ lb·ft/ft

Stabilizing moment is calculated as the sum of the moments of all weight components (wall footing, wall stem, and soil block above heel) about the toe:

$$M_s = \sum W_i x_i = 9 \times 3 \times 150 \times 4.5 + 12 \times 1 \times 150 \times 3.5 + 5 \times 12 \times 125 \times 6.5 = 73{,}275 \text{ lb·ft/ft}$$

$FS_{OT} = 73{,}275 / 19{,}898.4 = 3.7$

A

## 331

Cohesion = ½ × 1,600 = 800 psf

Slope angle = $\tan^{-1} 0.333 = 18.4°$

Assuming that no firm stratum exists within reasonable depth ("infinite" depth to firm stratum):

For $D \to \infty$ and slope inclination = 18.4°, stability number = 5.53

$$FS = \frac{Nc}{\gamma H} = \frac{5.53 \times 800}{125 \times 18} = 1.97$$

Correct answer is

D

## 332

Fraction passing no. 200 sieve, $F_{200} = 10$. Therefore, coarse fraction $= 90$. Since the coarse fraction is more than 50, the soil is predominantly coarse grained (first letter G or S). The cumulative retained on the no. 4 sieve, $R_4 = 100 - 80 = 20$ is less than half of the coarse fraction.

Therefore, first letter is S.

$LL = 31$, $PI = 31 - 25 = 6$

From the sieve data, we have the following critical sizes:

$D_{10} = 0.075$ mm; $D_{30} = 0.425$ mm; $D_{60} = 2.0$ mm

Uniformity coefficient: $C_u = \dfrac{D_{60}}{D_{10}} = \dfrac{2.0}{0.075} = 26.7$

Curvature coefficient: $C_z = \dfrac{D_{30}^2}{D_{60}D_{10}} = \dfrac{0.425^2}{2.0 \times 0.075} = 1.2$

$5 < F_{200} < 12$: Soil has dual classification. Meets criteria for SW ($C_u \geq 6$ and $1 < C_z < 3$) and plots below the A-line [$0.73(LL-20) = 8$; PI is less than 8], thereby second letter is M (silt) rather than C (clay).

Soil is classified as SW-SM. **D**

## 333

Footing size $B = 5$ ft. Normalizing with respect to footing size, we want the stress increase at 0.8B below and 1.2B lateral offset.

This point is on the stress contour $0.05p = 0.05 \times 100/25 = 0.2$ ksf $= 200$ psf **C**

## 334

Consider a unit width (1 ft) of the sheet piling. The load per unit length $= 500$ psf $\times$ 1 ft $= 500$ plf.

The sheet pile is considered to be hinged at the wales, and continuous over more than 3 wales, the design moment is given by:

$$M = \frac{wL^2}{10} = \frac{500 \times 4^2}{10} = 800 \text{ lb} \cdot \text{ft} = 9{,}600 \text{ lb} \cdot \text{in.}$$

Required section modulus: $S = \dfrac{M}{\sigma_{all}} = \dfrac{9,600}{1,400} = 6.86$ in.$^3$

For a rectangular section: $S = \dfrac{bt^2}{6} = 6.86 \Rightarrow t = \sqrt{\dfrac{6.86 \times 6}{12}} = 1.85$ in.          **D**

## 335

Unit weight of sand above GWT: $\gamma = 104 \times 1.05 = 109.2$ lb/ft$^3$

Unit weight of sand below GWT: $\gamma = 104 \times 1.18 = 122.7$ lb/ft$^3$

At the top of the clay layer, the effective stress: $\sigma'_v = 109.2 \times 6 + (122.7 - 62.4) \times 2 = 775.8$ lb/ft$^2$

Cyclic stress ratio: $\text{CSR} = \dfrac{\tau_{ave}}{\sigma'_v} \Rightarrow \tau_{ave} = 0.21 \times 775.8 = 162.9$ lb/ft$^2$

Shear strength of the saturated clay = cohesion $= \dfrac{1}{2} S_{uc} = 0.5 \times 2,000 = 1,000$ lb/ft$^2$

Factor of safety: $FS = \dfrac{1,000}{162.9} = 6.1$

Correct answer is          **D**

## 336

The component of the slice weight that acts *normal* to the failure surface is Wcos α. This creates the normal stress (σ) on that plane. The Mohr-Coulomb shear strength should be of the form

$c + \sigma \tan \phi = c + W\cos \alpha \tan \phi$

The component of the slice weight that acts *parallel* to the failure surface is Wsin α This causes the sliding effect along the failure plane. Therefore, this is the disturbing force and should be in the denominator of the expression for factor of safety. This eliminates **A** and **C**.

In **D**, the cohesion (a stabilizing force) is subtracted from the denominator rather than added to the numerator, as in **B**.

Correct answer is          **B**

## 337

Discharge $Q = 2,000$ gpm $= 4.464$ ft³/sec

Hydraulic conductivity $K = 1,000$ ft/day $= 1.157 \times 10^{-2}$ ft/sec

At observation well 1, radial distance $r_1 = 30$ ft, height of piezometric surface (above aquifer bottom): $y_1 = 50 - 4.5 = 45.5$ ft.

At observation well 2, radial distance $r_2 = 180$ ft, height of piezometric surface (above aquifer bottom): $y_2 = ?$

The steady-state equation for the drawdown due to pumping in an aquifer is:

$$Q = \frac{\pi K(y_1^2 - y_2^2)}{\ln\left(\frac{r_1}{r_2}\right)}$$

Here $y_1$ and $y_2$ are water table elevations at radial (horizontal) distance $r_1$ and $r_2$ (from pump well centerline), respectively.

$$Q = \frac{\pi K(y_1^2 - y_2^2)}{\ln\left(\frac{r_1}{r_2}\right)} \Rightarrow (y_1^2 - y_2^2) = \frac{Q\ln\left(\frac{r_1}{r_2}\right)}{\pi K} = \frac{4.464 \times \ln\left(\frac{30}{180}\right)}{\pi \times 1.157 \times 10^{-2}} = -220$$

Solving, we get $y_2 = 47.9$ ft. Drawdown $s_2 = 50 - 47.9 = 2.1$ ft.                    A

## 338

Since the GWT is within 5 ft (footing width) of the bottom of the footing, it must be accounted for.

In the third term of the bearing capacity equation, replace $\gamma$ with $\gamma_{ave}$:

$$\gamma_{ave} = \frac{\gamma D + (\gamma - \gamma_w)(B - D)}{B} = \frac{125 \times 4 + (125 - 62.4) \times (5 - 4)}{5} = 112.5$$

Using Terzaghi's bearing capacity factors: For $\phi = 35°$, $N_c = 58$, $N_q = 42$, $N_\gamma = 46$

Ultimate bearing capacity for a square footing:

$$q_{ult} = 1.3cN_c + \gamma DN_q + 0.4\gamma BN_\gamma = 0 + 125 \times 3 \times 42 + 0.4 \times 112.5 \times 5 \times 46 = 26,100 \text{ lb/ft}^2$$

Allowable bearing pressure: $q_{all} = \dfrac{q_{ult}}{FS} = \dfrac{26,100}{2.8} = 9,321 \text{ lb/ft}^2$

Allowable load: $Q_{all} = q_{all}A_f = 9,321 \times 25 = 233,025 \text{ lb} = 233 \text{ k}$

Correct answer is                                                                                                      A

## 339

Pile group dimensions 22 ft × 10 ft. End area, $A_{group} = 220 \text{ ft}^2$. Perimeter of "side-walls," $P_{group} = 2 \times (22 + 10) = 64$ ft.

The point bearing capacity of the pile group (in the clay layer) is approximated by:

$$Q_p = 9cA_{group} = 9 \times 1,400 \times 220 = 2,772,000 \text{ lb}$$

The side friction capacity (24-ft embedment in the clay layer) is given by:

$$Q_s = cLP_p = 1,400 \times 24 \times 64 = 2,150,400 \text{ lb}$$

Total pile group capacity = 2,772 + 2,150 = 4,922 k = 2,461 tons

Sum of individual capacities (of 15 piles) = 95 × 15 = 1,425 tons

Group efficiency = 2,461/1,425 = 1.73                                                                                   D

## 340

The depth associated with this SPT result is at the center of the 2nd and 3rd penetration intervals, that is, exactly 1 ft below the start depth for the test. Thus, the depth is 10 + 1 = 11 ft.

The effective vertical stress at $z = 11$ ft is:

$$\sigma'_v = 109.2 \times 6 + (122.7 - 62.4) \times 2 + (118 - 62.4) \times 3 = 942.6 \text{ psf} = 0.471 \text{ tsf}$$

Overburden correction factor (Liao & Whitman): $C_N = \sqrt{\dfrac{1}{\sigma'_v}} = \sqrt{\dfrac{1}{0.471}} = 1.46$

Corrected SPT N value: $N' = C_N N = 1.46 \times 39 = 57$                                                                 C

## Answer Key for Geotechnical Depth Exam

| | | | | | | | | | | |
|-----|---|---|-----|---|---|-----|---|---|-----|---|
| 301 | B | | 311 | C | | 321 | B | | 331 | D |
| 302 | D | | 312 | A | | 322 | C | | 332 | D |
| 303 | C | | 313 | D | | 323 | A | | 333 | C |
| 304 | C | | 314 | B | | 324 | D | | 334 | D |
| 305 | B | | 315 | B | | 325 | C | | 335 | D |
| 306 | A | | 316 | C | | 326 | A | | 336 | B |
| 307 | A | | 317 | B | | 327 | B | | 337 | A |
| 308 | C | | 318 | C | | 328 | D | | 338 | A |
| 309 | A | | 319 | B | | 329 | C | | 339 | D |
| 310 | A | | 320 | D | | 330 | A | | 340 | C |

# 12

# Water Resources & Environmental Depth Exam Solutions

These detailed solutions are for questions 401 to 440, representative of a 4-hour water resources & environmental depth exam according to the syllabus and guidelines for the Principles and Practice (P&P) of Civil Engineering Examination, administered by the National Council of Examiners for Engineering and Surveying (NCEES).

## 401

Among those listed in the water analysis, the contaminants affecting drinking water standards are: nitrate, turbidity, odor, total coliform, and total dissolved solids (TDS).

Nitrate concentration = 21 mg/L, converted to $NO_3$-N = $21 \times 14/62 = 4.74$ mg/L is within limits (10 mg/L) of primary DWS.

Turbidity exceeds limits of primary DWS (1.3 NTU > 1 NTU)

Odor is within limits (2.7 < 3.0 TON) of secondary DWS

Total coliform exceeds limits of primary DWS (1.7 MPN > 1.0 MPN)

TDS is within limits (437 < 500 mg/L) of secondary DWS

Therefore, turbidity and total coliform exceed drinking water standards.          D

## 402

| Time Period (hr) | Average Inflow Rate (ft³/sec) | Inflow (×10⁶ gal) | I–O |
|---|---|---|---|
| 00:00–02:00 | 18.3 | 0.986 | 0.407 |
| 02:00–04:00 | 12.4 | 0.668 | 0.089 |
| 04:00–06:00 | 8.7 | 0.469 | – |
| 06:00–08:00 | 7.3 | 0.393 | – |
| 08:00–10:00 | 6.4 | 0.345 | – |
| 10:00–12:00 | 8.9 | 0.479 | – |
| 12:00–14:00 | 14.5 | 0.781 | 0.202 |
| 14:00–16:00 | 18.9 | 1.018 | 0.439 |
| 16:00–18:00 | 6.5 | 0.350 | – |
| 18:00–20:00 | 5.6 | 0.302 | – |
| 20:00–22:00 | 8.3 | 0.447 | – |
| 22:00–24:00 | 13.2 | 0.711 | 0.132 |
| | | 6.949 | |

As an example, for the first period, Inflow $= Qt = 18.3$ cfs $\times 2$ hrs $= 18.3 \times 2 \times 3{,}600$ ft³ $= 985{,}565$ gal $= 0.986$ Mgal

Average flow per period $= 6.949/12 = 0.579$ Mgal

Longest streak of surplus intervals:

$(22{:}00 - 4{:}00) = 0.132 + 0.407 + 0.089 = 0.628$ Mgal                    B

## 403

Dissolved solids are those that pass through the filter in solution. The volatile dissolved solids will be those that are burnt off upon ignition.

In a 200-mL (0.2 L) solution: VDS $= 47.225 - 46.201 = 1.024$ g $= 1024$ mg

VDS concentration $= 1{,}024/0.2 = 5{,}120$ mg/L                    D

## 404

Let us assume that the factory can discharge a maximum lead concentration $= x$ (g/L)

Wastewater flow rate $= 2$ MGD $= 2 \times 1.5472 = 3.094$ ft³/sec

The lead concentration in the stream immediately after mixing:

$$\overline{Pb} = \frac{3.094x + 30 \times 4 \times 10^{-6}}{3.094 + 30}$$

Since the EPA limit on [Pb] is 15 µg/L:

$$\overline{Pb} = \frac{3.094x + 30 \times 4 \times 10^{-6}}{3.094 + 30} \leq 15 \times 10^{-6} \Rightarrow x \leq 121 \times 10^{-6}$$

Therefore, the plant must reduce its lead emission of 0.5 mg/L (500 µg/L) to 121 µg/L. This is a minimum removal rate of 76%.  A

## 405

Flow rate, $Q = 3,000$ gpm $= \dfrac{3,000}{448.8} = 6.684$ ft³/sec

For suction line:

Cross-section area: $A = \dfrac{\pi(1.5)^2}{4} = 1.767$ ft²

Velocity: $V = \dfrac{Q}{A} = \dfrac{6.684}{1.767} = 3.782$ ft/sec

Total head loss: $h_f + h_m = f\dfrac{L}{d}\dfrac{V^2}{2g} + \sum K\dfrac{V^2}{2g}$

$$h_f + h_m = 0.024 \times \frac{800}{1.5} \times \frac{3.782^2}{2 \times 32.2} + 5 \times \frac{3.782^2}{2 \times 32.2} = 3.95 \text{ ft}$$

For discharge line:

Cross-section area: $A = \dfrac{\pi(1.0)^2}{4} = 0.785$ ft²

Velocity: $V = \dfrac{Q}{A} = \dfrac{6.684}{0.785} = 8.51$ ft/sec

Total head loss: $h_f + h_m = 0.026 \times \dfrac{2,500}{1.0} \times \dfrac{8.51^2}{2 \times 32 \cdot 2} + 25 \times \dfrac{8.51^2}{2 \times 32.2} = 101.2$ ft

Total dynamic head = static head + head loss = 80 + 3.95 + 101.2 = 185.15 ft

Pump power rating:

$$P = \frac{\gamma QH}{\eta} = \frac{62.4 \times 6.684 \times 185.15}{0.88} = 87,753 \text{ lb} \cdot \text{ft/sec} = 159.6 \text{ hp}$$  C

## 406

As the end contractions are not suppressed, the effective width of the weir opening is:

$$b_{eff} = b - 0.1nH = 6 - 0.1 \times 2 \times 5.4 = 4.92 \text{ ft}$$

Weir discharge is calculated from:

$$Q = 3.33 b_{eff} H^{3/2} = 3.33 \times 4.92 \times 5.4^{3/2} = 205.6 \text{ ft}^3/\text{sec} \qquad \text{D}$$

## 407

The simplest method for determining base flow is the "constant base flow" approach, which seems appropriate (because the discharge at $t = 0$ and $t = 6$ are nearly equal). After the base-flow ($Q_b = 23$) is subtracted, we get the pattern of net discharge caused by storm runoff:

| Time (hr) | 0 | 1 | 2 | 3 | 4 | 5 | 6 |
|---|---|---|---|---|---|---|---|
| Discharge Q (ft³/sec) | 0 | 61 | 104 | 89 | 52 | 9 | 2 |

Volume of runoff: $V = \Sigma Q \times \Delta t = 317 \times 3,600 = 1,141,200 \text{ ft}^3$

Average depth of runoff (excess precipitation) = $1,141,200/(115 \times 43,560) = 0.228 \text{ ft} = 2.73 \text{ in.}$

The peak *runoff* discharge (ft³/sec) due to a 2-hr storm producing 1.7 in. of runoff is most nearly $104 \times 1.7/2.73 = 65 \text{ ft}^3/\text{sec}$.

Adding in the base flow (unaffected by storm), the corresponding peak discharge recorded in stream = $65 + 23 = 88 \text{ ft}^3/\text{sec}$. A

## 408

The following values of curve number are obtained from TR55 (Urban Hydrology for Small Watersheds).

| Region | Area (acres) | Soil Type | Land Use | CN |
|--------|--------------|-----------|----------|-----|
| 1 | 80 | C | Single family homes on ½-acre lots | 80 |
| 2 | 50 | D | Lawns in good condition | 80 |
| 3 | 10 | B | Paved streets and sidewalks | 98 |
| 4 | 50 | C | Grassy areas: fair condition | 79 |
| 5 | 40 | A | Woods: fair condition | 36 |

Composite curve number:

$$\overline{CN} = \frac{\sum CN_i \times A_i}{\sum A_i} = \frac{80 \times 80 + 80 \times 50 + 98 \times 10 + 79 \times 50 + 36 \times 40}{230} = 72.9 \qquad \text{C}$$

## 409

According to the Darcy-Weisbach model, frictional head loss in a circular pipe segment (length $L$, diameter $D$, flow rate $Q$) is given by:

$$h_f = \left( \frac{0.0311 fL}{D^5} \right) Q^2 = kQ^2 \qquad (L \text{ in ft}, D \text{ in in.}, Q \text{ in gpm})$$

The Hardy-Cross (iterative) method uses the following correction to the presumed flow rate:

$$\Delta Q = -\frac{\sum kQ^2}{\sum |2kQ|}$$

For each pipe, the parameter $k$ is calculated and shown in the table. The units-related constant (0.0311) shows up in numerator and denominator and is left out for convenience.

| Pipe | Length (ft) | Friction Factor | Diameter (in.) | $k = fL/D^5 \times 10^4$ |
|------|-------------|-----------------|----------------|--------------------------|
| AB | 600 | 0.020 | 8 | 3.662 |
| AC | 700 | 0.030 | 12 | 0.844 |
| AD | 500 | 0.020 | 6 | 12.86 |
| CD | 400 | 0.025 | 8 | 3.052 |
| BC | 300 | 0.030 | 12 | 0.362 |

*(Continued on next page)*

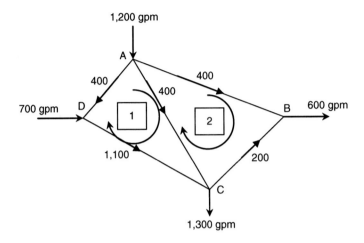

1,200 gpm

For loop 1 (clockwise: AC-CD-DA), the correction to the discharge pattern is calculated as:

$$\Delta Q = -\frac{\sum kQ^2}{\sum |2kQ|} = -\frac{0.844 \times 400^2 - 3.05 \times 1{,}100^2 - 12.86 \times 400^2}{2 \times 0.844 \times 400 + 2 \times 3.05 \times 1{,}100 + 2 \times 12.86 \times 400} = 317.7 \text{ gpm}$$

$Q_{AC} = 400 + 318 = 718$ gpm

$Q_{DC} = 1100 - 318 = 782$ gpm

$Q_{AD} = 400 - 318 = 82$ gpm (A to D)

A

## 410

Using the Hazen-Williams equation for head loss, the total friction head loss is given by:

$$h_f = \frac{10.429 \times Q_{gpm}^{1.85} \times L_{ft}}{C^{1.85} \times D_{in}^{4.865}}$$

$$h_f = \frac{10.429 \times Q_{gpm}^{1.85} \times 2{,}000}{100^{1.85} \times 18_{in}^{4.865}} + \frac{10.429 \times Q_{gpm}^{1.85} \times 3{,}000}{100^{1.85} \times 12_{in}^{4.865}} = 3.834 \times 10^{-5} \times Q_{gpm}^{1.85}$$

At $Q = 1{,}000$ gpm, $h_f = 13.6$ ft + 10% minor loss = 15 ft, total dynamic head = $70 + 15 = 85$ ft

At $Q = 1{,}400$ gpm, $h_f = 25.4$ ft + 10% minor loss = 28 ft, total dynamic head = $70 + 28 = 98$ ft

At $Q = 1,800$ gpm, $h_f = 40$ ft + 10% minor loss = 44 ft, total dynamic head = $70 + 44 = 114$ ft

The system curve is plotted with coordinates $(Q, H) = (0, 70)$, $(1,000, 85)$, $(1,400, 98)$, $(1,800, 114)$.

The H-W curve for the pumps in series is obtained by doubling the ordinates of the given pump curve. The operating point (intersection) seems to be at $Q = 1,050$ gpm.

B

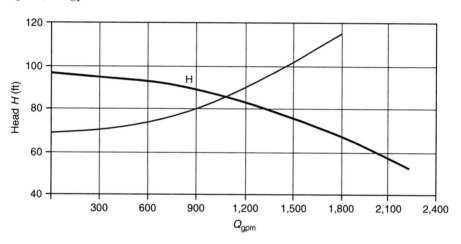

## 411

Surface loading velocity $V = 10$ ft/hr = 0.002778 ft/sec

Incident flow rate $Q = 3.2$ MGD = $3.2 \times 1.5472$ ft³/sec

Based on hydraulic load, total area of filter needed:
$A = Q/V = (3.2 \times 1.5472)/0.002778 = 1,782.4$ ft²

Solid load in influent: $X = 800$ mg/L $\times 3.2$ MGD $\times 8.3454 = 21,364$ lb/day

Based on solid load, total area of filter needed: $A = 21,364/15 = 1,424.3$ ft²

The other criterion (1,782.4 ft²) governs.

Number of filters needed = $1,782.4/250 = 7.13$

At least 8 filters are needed.

C

## 412

At 5 days, the BOD is expected to be on the exponential BOD growth curve. Therefore, the ultimate BOD is calculated from:

$$BOD_t = BOD_{ult}(1-10^{-kt}) \Rightarrow BOD_{ult} = 4.5/(1-10^{-0.1\times5}) = 6.58 \text{ mg/L}$$

According to the same model, the 20-day carbonaceous BOD would have been:

$$BOD_{20} = 6.58\times(1-10^{-0.1\times20}) = 6.52 \text{ mg/L}$$

At 20 days, the excess (recorded) BOD comes from nitrogenous bacteria.

Nitrogenous BOD = 8.3 − 6.52 = 1.78 mg/L               **D**

## 413

Assuming average inflow during 5-min analysis period = 50 ft$^3$/sec

During this stage, head at broad crested weir = 126.4 − 125.0 = 1.4 ft.

*Note:* This is the weir head at the beginning of the analysis period and this rough solution assumes this stays constant, even though it does not. A better solution would be to break down the analysis period into smaller intervals.

Effective width of (rectangular) weir: $b_e = b - 0.1NH = 5 - 0.1\times2\times1.4 = 4.72$ ft

Assuming weir coefficient = 3.3, average outflow during this stage:
$Q = CbH^{3/2} = 3.3\times4.72\times1.4^{1.5} = 25.8$ cfs

Net inflow during analysis period (5 min) = 50 − 25.8 = 24.2 cfs

Net storage volume increase into pond during this period (5 min) = 24.2 × 5 × 60 = 7,260 ft$^3$

Depth increase = 7,260/43,560 = 0.167 ft

Pool elevation = 126.4 + 0.167 = 126.567 ft

Weir head = 126.567 − 125.0 = 1.567 ft

Weir discharge $Q = CbH^{3/2} = 3.3\times4.72\times1.567^{1.5} = 30.6$ cfs               **A**

## 414

Orifice diameter, $A_0 = \dfrac{\pi}{4}\left(\dfrac{2}{12}\right)^2 = 0.0218\ \text{ft}^2$

Pipe diameter, $A_1 = \dfrac{\pi}{4}\left(\dfrac{4}{12}\right)^2 = 0.0873\ \text{ft}^2$

Discharge coefficient of an orifice meter is given by:

$$C_f = \frac{C_v C_c}{\sqrt{1 - C_c^2 A_0^2 / A_1^2}} = \frac{0.95 \times 0.9}{\sqrt{1 - 0.9^2 \times 0.0218^2 / 0.0873^2}} = 0.878$$

Pressure difference = 30 psi = 30 × 144 psf

$$Q = C_f A_0 \sqrt{\frac{2g\Delta p}{\gamma}} = 0.878 \times 0.0218 \times \sqrt{\frac{2 \times 32.2 \times 30 \times 144}{62.4}} = 1.28\ \text{cfs} = 574\ \text{gal/min}$$

**B**

## 415

Hydraulic gradient: $i = \dfrac{H}{L} = \dfrac{945 - 905}{160} = 0.25\ \text{ft/ft}$

Porosity: $n = \dfrac{e}{1+e} = \dfrac{0.45}{1+0.45} = 0.31$

Seepage velocity: $v_s = \dfrac{Ki}{n} = \dfrac{0.25 \times 1 \times 10^{-4}}{0.31} = 8.1 \times 10^{-5}\ \text{ft/sec}$

Scour velocity = 8 in./hr = $1.85 \times 10^{-4}$ ft/sec

$$FS = \frac{1.85 \times 10^{-4}}{8.1 \times 10^{-5}} = 2.29$$

**C**

## 416

The head difference (top of first sand layer to bottom of clay layer) = 18 in.

Hydraulic gradient $= \dfrac{18\ \text{in.}}{12\ \text{in.}} = 1.5\ \text{in./in.}$

(*Continued on next page*)

Porosity: $n = \dfrac{e}{1+e} = \dfrac{0.7}{1.7} = 0.412$

Seepage velocity: $V_s = \dfrac{Ki}{n} = \dfrac{1.5 \times 1.2 \times 10^{-5}}{0.412} = 4.369 \times 10^{-5}\,\text{ft/hr} = 0.0126\,\text{in./day}$

Time for leachate to seep through the 12-in. layer $= \dfrac{12}{0.0126} = 954$ days ⟶ A

## 417

For pipe diameter $= 36$ in. and $Q = 100$ cfs, FHWA nomograph yields $HW/D = 3.5$

Headwater $= 3.5 \times 3 = 10.5$ ft (this is measured from the upstream invert elevation)

100-year water surface elevation $= 325 + 10.5 = 335.5$ ft

Vertical clearance between roadway and 100-year WSE $= 337.8 - 335.5 = 2.3$ ft ⟶ B

## 418

No matter how the flow distributes through the network, the head loss is proportional to $Q^2$ (if Darcy-Weisbach model is used) or $Q^{1.85}$ (if Hazen-Williams model is used). When the inflow changes from 300 gpm to 500 gpm, the (total) head loss becomes:

$h_f = 70 \times (500/300)^2 = 194.4$ ft    (180 ft if Hazen-Williams model is used)

As 1 atm $= 14.7$ psi $= 33.9$ ft of water, 194.4 ft of head loss $= 84.3$ psi (78.3 psi if Hazen-Williams model is used). ⟶ D

## 419

For masonry construction, $F = 0.8$

Fire flow demand: $Q = 18F\sqrt{A} = 18 \times 0.8 \times \sqrt{160{,}000} = 5{,}760$ gpm

Time for which the tank can provide this flow rate: $t = V/Q = 700{,}000/5{,}760 = 121.5$ min

Therefore, the tank meets fire flow demand for approximately 2 hrs. ⟶ C

## 420

During the 3 years of construction, the volume of sediment captured by the pond:

$$V = 1,200 \times 280 \times 3 = 1,008,000 \text{ ft}^3$$

Depth of sediment in pond during first 3 years:

$$d = V/A = (1.008 \times 10^6)/(5 \times 43,560) = 4.63 \text{ ft}$$

Original depth of water in pond = 353.4 − 345.6 = 7.8 ft

Allowable depth of sediment (before sediment removal) = 7.8 − 3.0 = 4.8 ft

Given that 4.63 ft of sediment accumulates in first 3 years, remaining depth (before dredging) = 0.17 ft. This corresponds to a volume = $0.17 \times 5 \times 43,560 = 37026 \text{ ft}^3$.

Additional time to accumulate this volume of sediment = $t = 37,026/(280 \times 300) = 0.44$ year

Number of years until sediment must be removed from pond = 3.44 years          **B**

## 421

Flow rate $Q = 500 \text{ gpm} = \dfrac{500}{448.8} = 1.114 \text{ ft}^3/\text{sec}$

Nominal density of water $= \dfrac{62.4}{32.2} = 1.938 \text{ lb-sec}^2/\text{ft}$

Initial velocity (out of nozzle): $V = \dfrac{Q}{\dfrac{\pi}{4}D^2} = \dfrac{1.114}{\dfrac{\pi}{4} \times \left(\dfrac{2}{12}\right)^2} = 51.1 \text{ ft/sec}$

Final velocity (after hitting deflector) = $V\cos 40i + V\sin 40j = 39.1i + 32.8j$

Change in velocity = $\Delta V = 39.1i + 32.8j - 51.1i = -12i + 32.8j$

Force, $F = \rho Q \Delta V = 1.938 \times 1.114 \times (-12i + 32.8j) = 25.91i + 70.82j$

Resultant force magnitude: $F = \sqrt{F_x^2 + F_y^2} = \sqrt{25.91^2 + 70.82^2} = 75.4 \text{ lb}$          **C**

## 422

Time for stream to travel 20 miles (105,600 ft) at 4 ft/sec is 26,400 sec (7.33 hr).

Assuming that the stream returns to its equilibrium temperature of 12°C within a short distance, there is no need to calculate the weighted average temperature at the instant of mixing.

At temperature of 12°C, saturation D.O. = 10.77 mg/L (see Table 305.5 in I. Goswami, *Civil Engineering PE All-in-One Exam Guide: Breadth and Depth*, McGraw-Hill, 2009.)

At a distance of 20 mi (7.33 hr away), the dissolved oxygen concentration = 4.4 mg/L (from the given sag curve)

Therefore, 5 mi downstream, the oxygen deficit = 10.77 − 4.4 = 6.37 mg/L          **B**

## 423

This is an "unknown-depth-of-flow" problem. It is best solved using the tables (see *All-in-One* Table 303.3 and Equation 303.37).

Flow parameter: $K = \dfrac{Qn}{kb^{8/3}S^{1/2}} = \dfrac{150 \times 0.016}{1.486 \times 10^{8/3} \times 0.004^{1/2}} = 0.055$

For the $m = 2$ (side slope parameter), the fragment of the table is shown below:

| d/b | 0 | 0.25 | 0.5 | 0.75 | 1.0 | 1.5 | 2.0 | 2.5 | 3.0 | 3.5 | 4.0 |
|---|---|---|---|---|---|---|---|---|---|---|---|
| 0.01 | 0.0005 | 0.0005 | 0.0005 | 0.0005 | 0.0005 | 0.0005 | 0.0005 | 0.0005 | 0.0005 | 0.0005 | 0.0005 |
| 0.02 | 0.0014 | 0.0014 | 0.0015 | 0.0015 | 0.0015 | 0.0015 | 0.0015 | 0.0015 | 0.0015 | 0.0015 | 0.0015 |
| 0.04 | 0.0044 | 0.0045 | 0.0046 | 0.0046 | 0.0047 | 0.0047 | 0.0048 | 0.0048 | 0.0049 | 0.0049 | 0.0050 |
| 0.06 | 0.0085 | 0.0087 | 0.0089 | 0.0090 | 0.0091 | 0.0093 | 0.0095 | 0.0096 | 0.0098 | 0.0099 | 0.0101 |
| 0.08 | 0.0135 | 0.0139 | 0.0142 | 0.0145 | 0.0147 | 0.0152 | 0.0155 | 0.0159 | 0.0162 | 0.0165 | 0.0168 |
| 0.1 | 0.0191 | 0.0198 | 0.0204 | 0.0209 | 0.0214 | 0.0221 | 0.0228 | 0.0234 | 0.0241 | 0.0247 | 0.0253 |
| 0.12 | 0.0253 | 0.0265 | 0.0275 | 0.0283 | 0.0290 | 0.0303 | 0.0314 | 0.0324 | 0.0334 | 0.0345 | 0.0355 |
| 0.14 | 0.0320 | 0.0338 | 0.0352 | 0.0365 | 0.0376 | 0.0395 | 0.0412 | 0.0428 | 0.0444 | 0.0459 | 0.0475 |
| 0.16 | 0.0392 | 0.0416 | 0.0437 | 0.0455 | 0.0471 | 0.0498 | 0.0523 | 0.0546 | 0.0569 | 0.0591 | 0.0614 |
| 0.18 | 0.0467 | 0.0500 | 0.0529 | 0.0553 | 0.0575 | 0.0612 | 0.0646 | 0.0678 | 0.0710 | 0.0741 | 0.0772 |
| 0.2 | 0.0547 | 0.0589 | 0.0627 | 0.0659 | 0.0687 | 0.0737 | 0.0783 | 0.0826 | 0.0868 | 0.0910 | 0.0952 |
| 0.22 | 0.0629 | 0.0683 | 0.0731 | 0.0772 | 0.0809 | 0.0874 | 0.0932 | 0.0989 | 0.1043 | 0.1098 | 0.1152 |
| 0.24 | 0.0714 | 0.0781 | 0.0841 | 0.0893 | 0.0939 | 0.1021 | 0.1096 | 0.1167 | 0.1237 | 0.1306 | 0.1374 |

Table header: **Horizontal projection** $m$

Interpolating, $d/b = 0.164$

$d = 1.64$ ft = 19.7 in.          **D**

## 424

Treating this as a broad-crested weir, the weir coefficient may be found (table shown). For crest breadth (measured parallel to flow) = 12 in. = 1 ft and measured head = 9 in. = 0.75 ft, the weir coefficient $C = 2.82$.

| Measured head (ft) | Weir crest breadth w (ft) | | | | | | | | | | |
|---|---|---|---|---|---|---|---|---|---|---|---|
| | 0.5 | 0.75 | 1.0 | 1.5 | 2.0 | 2.5 | 3.0 | 4.0 | 5.0 | 10.0 | 15.0 |
| 0.2 | 2.80 | 2.75 | 2.69 | 2.62 | 2.54 | 2.48 | 2.44 | 2.38 | 2.34 | 2.49 | 2.68 |
| 0.4 | 2.92 | 2.80 | 2.72 | 2.64 | 2.61 | 2.60 | 2.58 | 2.54 | 2.50 | 2.56 | 2.70 |
| 0.6 | 3.08 | 2.89 | 2.75 | 2.64 | 2.61 | 2.60 | 2.68 | 2.69 | 2.70 | 2.70 | 2.70 |
| 0.8 | 3.30 | 3.04 | 2.85 | 2.68 | 2.60 | 2.60 | 2.67 | 2.68 | 2.68 | 2.69 | 2.64 |
| 1.0 | 3.32 | 3.14 | 2.98 | 2.75 | 2.66 | 2.64 | 2.65 | 2.67 | 2.68 | 2.68 | 2.63 |
| 1.2 | 3.32 | 3.20 | 3.08 | 2.86 | 2.70 | 2.65 | 2.64 | 2.67 | 2.66 | 2.69 | 2.64 |
| 1.4 | 3.32 | 3.26 | 3.20 | 2.92 | 2.77 | 2.68 | 2.64 | 2.65 | 2.65 | 2.67 | 2.64 |
| 1.6 | 3.32 | 3.29 | 3.28 | 3.07 | 2.89 | 2.75 | 2.68 | 2.66 | 2.65 | 2.64 | 2.63 |

Flow rate: $Q = CbH^{3/2}$

Flow rate per unit width: $Q/b = CH^{3/2} = 2.82 \times 0.75^{3/2} = 1.83$ cfs/ft

**C**

## 425

Let us assume that the volume carried by the channel itself is negligible. The floodplain may then be modeled as a wide rectangular channel (depth << width) whose width shrinks from 800 ft to 300 ft as a result of development.

For a wide rectangular channel, hydraulic radius: $R_h$ = Area/Perimeter = $wd/(w+2d) \approx d$

Flow rate: $Q = VA = \left( \dfrac{1.486}{n} R_h^{2/3} S^{1/2} \right) A \approx \left( \dfrac{1.486}{n} d^{2/3} S^{1/2} \right) wd = Cwd^{5/3}$

Therefore, for the same flood flow rate, the product $wd^{5/3}$ is a constant.

The original depth of flow = 252.7 − 250.95 = 1.75 ft

Original width = 800 ft

New width = 300 ft

$w_1 d_1^{5/3} = w_2 d_2^{5/3} \Rightarrow d_2 = \left( \dfrac{w_1 d_1^{5/3}}{w_2} \right)^{3/5} = \left( \dfrac{800 \times 1.75^{5/3}}{300} \right)^{3/5} = 3.15$ ft

New flood elevation = 250.95 + 3.15 = 254.1 ft

**B**

## 426

Assuming that the primary sludge, which contains 6% solids, essentially has the same specific gravity as water, we can convert the 6% solids to a concentration:

Solids 6% (by weight) = 60 g/1,000 g = 60,000 mg/L (assuming sludge is "watery," 1 L weighs 1 kg = 1,000 g)

Primary effluent contains TSS = $0.35 \times 250 = 87.5$ mg/L

Influent contains 250 mg/L TSS. After 65% removal, primary effluent contains $0.35 \times 250 = 87.5$ mg/L TSS.

Performing a summation of mass flow rates at the node representing the primary clarifier, we get (where $Q_{ps}$ is the volumetric flow rate of primary sludge in MGD):

$$4 \times 250 = (4 - Q_{ps}) \times 87.5 + Q_{ps} \times 60,000 \Rightarrow Q_{ps} = 0.01085 \text{ MGD} = 10,850 \text{ gpd}$$    **C**

## 427

The "as species" concentrations of the divalent metal ions ($Ca^{++}$, $Mg^{++}$, and $Fe^{++}$) should be converted to "as $CaCO_3$" equivalents utilizing the factors calculated from equivalent weight.

$[Ca^{++}] = 60.0 \times 2.5 = 150$ mg/L as $CaCO_3$

$[Mg^{++}] = 21.2 \times 4.1 = 86.9$ mg/L as $CaCO_3$

$[Fe^{++}] = 2.2 \times 1.77 = 3.9$ mg/L as $CaCO_3$

Hardness = 240.8 mg/L as $CaCO_3$    **A**

## 428

For a rectangular open channel, critical velocity is given by: $V_c = \sqrt{gd_c} = 15$

Therefore, critical depth: $d_c = 15^2/32.2 = 6.99$ ft

Flow area = $6.99 \times 10 = 69.9$ ft$^2$

Flow rate: $Q = VA = 15 \times 69.9 = 1,048$ ft$^3$/sec    **B**

## 429

Suction line: 800-ft length, 18-in. diameter, $C = 100$

Discharge line: 2,500-ft length, 12-in. diameter, $C = 100$

Frictional loss is given by: $h_f = \dfrac{10.429 \times Q_{gpm}^{1.85} \times L_{ft}}{C^{1.85} \times d_{in.}^{4.865}}$

Total head loss:

$$h_f = \frac{10.429 \times 3,000^{1.85} \times 800}{100^{1.85} \times 18^{4.865}} + \frac{10.429 \times 3,000^{1.85} \times 2,500}{100^{1.85} \times 12^{4.865}} = 3.52 + 79.18 = 82.7$$

Total dynamic head = static head + head loss = 162.7 ft

The fluid power: $P = \gamma QH = 62.4 \times 6.7 \times 162.7 = 67,985 \ \text{lb} \cdot \text{ft/sec} = 123.6 \ \text{hp}$

Brake horsepower = 123.6/0.85 = 145 hp          **D**

## 430

$Q = 120,000 \ \text{gpm} = 268 \ \text{ft}^3/\text{sec}$

Flow per unit width: $q = \dfrac{268}{15} = 17.87 \ \text{ft}^2/\text{sec}$

Critical depth: $d_c = \left(\dfrac{q^2}{g}\right)^{1/3} = \left(\dfrac{17.87^2}{32.2}\right)^{1/3} = 2.15 \ \text{ft}$

Therefore, the depth of 15 in. (1.25 ft) is supercritical. A hydraulic jump occurs at this location as long as the tailwater depth $d_2$ corresponds to $d_1 = 1.25$.

The velocity at the bottom of the spillway is: $V_1 = Q/bd = 268/(15 \times 1.25) = 14.3 \ \text{ft/sec}$

The tailwater depth $d_2$ which is the conjugate depth of depth $d_1$ is given by:

$$d_2 = -\frac{1}{2}d_1 + \sqrt{\frac{2V_1^2 d_1}{g} + \frac{d_1^2}{4}} = -\frac{1.25}{2} + \sqrt{\frac{2 \times 14.3^2 \times 1.25}{32.2} + \frac{1.25^2}{4}} = 3.41 \ \text{ft} \qquad \textbf{B}$$

## 431

Using the NRCS approach, we can calculate the runoff depth.

The composite curve number is given by:

$$\overline{CN} = \frac{\sum CN_i A_i}{\sum A_i} = \frac{69 \times 80 + 45 \times 80 + 98 \times 50 + 87 \times 90 + 35 \times 70}{370} = 66$$

For $CN = 66$ and a gross rainfall $P_g = 5.6$ in.

Soil storage capacity, $S = \dfrac{1{,}000}{66} - 10 = 5.15$

Runoff depth, $Q = \dfrac{(P_g - 0.2S)^2}{P_g + 0.8S} = \dfrac{(5.6 - 0.2 \times 5.15)^2}{5.6 + 0.8 \times 5.15} = 2.15$ in.    **D**

The TR55 documentation also has a figure that can be used to look up runoff depth $Q$ for a given value of gross rainfall $P_g$ and curve number $CN$.

## 432

If the bypass factor is $x$, then the fraction $1 - x$ gets treatment (and removal), while the fraction $x$ gets no removal.

Incoming hardness quantity − Transmitted hardness = Removed hardness

$$Q \times 200 - Q \times 50 = 0.88 \times Q \times (1 - x) \times 200 \Rightarrow x = 0.15$$

Bypass fraction = 15%    **B**

## 433

Time of concentration = time for sheet flow + time for channel flow = $(200/0.5) + (400/2.0) = 600$ sec = 10 min

Using Steel formula (for Reno, NV, which is in region 7), for 50-year recurrence interval, coefficients are $K = 65$, $b = 8$.

Design rainfall intensity, $I = K/(t_c + b) = 65/(10 + 8) = 3.61$ in./hr

For paved surfaces, use a rational coefficient = 0.98. Area = $400 \times 400 = 160{,}000$ ft$^2$ = 3.673 acres

Runoff: $Q = CIA_d = 0.98 \times 3.61 \times 3.673 = 12.99$ acre · in./hr $= 13.1$ ft$^3$/sec    **A**

## 434

For a perpetual annuity, capitalized cost is $A/i$. All costs are calculated in thousands of dollars.

Present worth of Plan A (single capital investment) is calculated as the sum of initial expense + present worth of perpetual annuity:

$$P_A = 420 + 40 \times \frac{1}{0.07} = 991.43$$

Present worth of Plan B (two-stage capital investment) is calculated as the sum of initial expense + lump-sum expense at $t = 10$ years + perpetual annuity of 16 + 10 year annuity of 4:

$$P_A = 200 + 320 \left( \frac{P}{F}, 10 \text{ years}, 7\% \right) + 16 \times \frac{1}{0.07} + 4 \left( \frac{P}{A}, 10 \text{ years}, 7\% \right)$$

$$= 200 + 320 \times 0.5083 + \frac{16}{0.07} + 4 \times 7.0236 = 619.322$$

Cost ratio = 619.322/991.43 = 0.625                                                  **A**

## 435

Volume of trickling filter: $V = \dfrac{\pi}{4} \times 80^2 \times 6 = 30{,}159 \text{ ft}^3$

BOD load = $240 \times 3 \times 8.3454 = 6{,}008.7$ lb/day

BOD loading rate: $L_{BOD} = \dfrac{6{,}008.7}{30.159} = 199.2 \dfrac{\text{lb}}{1{,}000 \text{ ft}^3 - \text{day}}$

Recirculation factor: $F = \dfrac{1 + R}{(1 + 0.1R)^2} = \dfrac{1 + 3}{(1 + 0.3)^2} = 2.37$

Removal efficiency: $\eta = \dfrac{1}{1 + 0.0561\sqrt{\dfrac{199.2}{2.37}}} = 0.66$

Therefore, the $BOD_5$ in the effluent = 34% of 240 = 81.6 mg/L                      **B**

## 436

Discharge $Q = 2,000$ gpm $= 4.464$ ft³/sec

Hydraulic conductivity $K = 1,000$ ft/day $= 1.157 \times 10^{-2}$ ft/sec

At observation well 1, radial distance $r_1 = 30$ ft, height of piezometric surface (above aquifer bottom): $y_1 = 50 - 4.5 = 45.5$ ft

At observation well 2, radial distance $r_2 = 180$ ft, height of piezometric surface (above aquifer bottom): $y_2 = ?$

$$Q = \frac{\pi K (y_1^2 - y_2^2)}{\ln\left(\dfrac{r_1}{r_2}\right)} \Rightarrow (y_1^2 - y_2^2) = \frac{Q \ln\left(\dfrac{r_1}{r_2}\right)}{\pi K} = \frac{4.464 \times \ln\left(\dfrac{30}{180}\right)}{\pi \times 1.157 \times 10^{-2}} = -220$$

$y_2 = 47.9$ ft

Drawdown $s_2 = 50 - 47.9 = 2.1$ ft                                        A

## 437

Influent TSS $= 450$ mg/L

Influent VSS $= 0.65 \times 450 = 292.5$ mg/L

Digester removes 60% of VSS $= 0.6 \times 292.5 = 175.5$ mg/L

Therefore, digester transmits TSS $= 450 - 175.5 = 274.5$ mg/L

Total suspended solids load in the effluent $= 274.5 \times 3 \times 8.3454 = 6,872$ lb/day        C

**438**

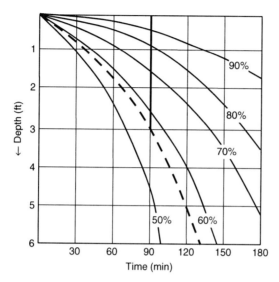

It seems that the iso-concentration line passing through the point of interest (3 ft, 90 min) corresponds to approximately 57% removal. This is shown by a dashed curve. The heavy vertical line extending to the top edge of the curve can be used to conduct the following weighted average.

$$\%R = \frac{57+60}{2} \times \frac{0.4}{3.0} + \frac{60+70}{2} \times \frac{1.0}{3.0} + \frac{70+80}{2} \times \frac{0.8}{3.0} + \frac{80+90}{2} \times \frac{0.4}{3.0} +$$

$$\frac{90+100}{2} \times \frac{0.4}{3.0} = 73.5$$

Cumulative removal at depth of 3 ft at 90 min = 74%

D

**439**

Chlorine residual = dose − demand = 5.0 − 3.5 = 1.5 mg/L

At this concentration, for 2-log inactivation, $CT = 68$ mg/L-min

Therefore, contact time $T = 68/1.5 = 45.33$ min = 0.0315 day

Volume of chlorination chamber:

$V = QT = 3.5 \times 10^6 \times 0.0315 = 14{,}730$ gal = 14,730 ft$^3$

A

## 440

VSS concentration in wastewater $= 0.75 \times 540 = 405$ mg/L

VSS load in wastewater $= 405 \times 4 \times 8.3454 = 13{,}519.5$ lb/day $= 563.3$ lb/hr

Volume required (based on volatile solids load) $= 563.3/0.0021 = 268{,}238$ gal

Volume required (based on hydraulic detention time) $= Qt = 4 \times 10^6 \times 2/24 = 333{,}333$ gal

Therefore, required volume $= 333{,}333$ gal $= 44{,}563$ ft$^3$

Total volume (ft$^3$) of choices **A** through **D** are: 33,928, 24,126, 50,894, 40,032, respectively.

Correct answer is                                                                                  C

# Answer Key for Water Resources & Environmental Depth Exam

| | | | | | | | | |
|---|---|---|---|---|---|---|---|
| 401 | D | 411 | C | 421 | C | 431 | D |
| 402 | B | 412 | D | 422 | B | 432 | B |
| 403 | D | 413 | A | 423 | D | 433 | A |
| 404 | A | 414 | B | 424 | C | 434 | A |
| 405 | C | 415 | C | 425 | B | 435 | B |
| 406 | D | 416 | A | 426 | C | 436 | A |
| 407 | A | 417 | B | 427 | A | 437 | C |
| 408 | C | 418 | D | 428 | B | 438 | D |
| 409 | A | 419 | C | 429 | D | 439 | A |
| 410 | B | 420 | B | 430 | B | 440 | C |

# 13

# Transportation Depth Exam Solutions

These detailed solutions are for questions 501 to 540, representative of a 4-hour transportation depth exam according to the syllabus and guidelines for the Principles and Practice (P&P) of Civil Engineering Examination, administered by the National Council of Examiners for Engineering and Surveying (NCEES).

## 501

The back tangent has bearing N 46°25'32"E. Given that this is in the NE quadrant, this is also its azimuth. The forward tangent has bearing S 17°56'21"E. As this is in the SE quadrant, its azimuth = 180° − 17°56'21" = 162°3'39".

The deflection angle is the difference of the azimuths = 162°3'39" − 46°25'32" = 115°38'07" (115.635°).

Length of back tangent $T_a$ = distance from PC to PI = 15,812.98 − 13,834.12 = 1,978.86 ft

$$T_a = \frac{R_2 - R_1 \cos I + (R_1 - R_2)\cos\Delta_2}{\sin I}$$

(Continued on next page)

$$1,978.86 = \frac{900 - 1,800 \times \cos 115.635 + (1,800 - 900)\cos\Delta_2}{\sin 115.635} \Rightarrow \Delta_2 = 83.279,$$

$$\Delta_1 = 32.356$$

$$L_1 = 100\frac{\Delta_1}{D_1} = 100\frac{\Delta_1 R_1}{5,729.578} = 1,016.49 \text{ ft}$$

$$\text{PCC} = \text{PC} + L_1 = 138 + 34.12 + 10 + 16.49 = 148 + 50.61 \qquad \text{C}$$

## 502

Distance from $\text{PVC}_2$ to $\text{PVI}_2 = 42.7213 - 35.2386 = 7.4827$ sta. This represents half the length of curve. Therefore, length of curve 2 = 14.9654 sta.

Rate of gradient change for curve 2, $R = \dfrac{9}{14.9654} = 0.60139$ %/sta

Elevation of $\text{PVC}_2 = \text{elev PVI}_2 - G_1\dfrac{L}{2} = 374.52 - (-5) \times 7.4827 = 411.9$ ft

Distance to bridge structure from $\text{PVC}_2 = 45.55 - 35.2386 = 10.3114$ sta

Curve elevation at 40 + 55 is:

$$y = 411.93 + (-5)\times10.3114 + \frac{1}{2}\times0.60139\times10.3114^2 = 392.34 \text{ ft}$$

Vertical clearance = 405.54 − 392.34 = 13.20 ft

For a sag curve with limited clearance, AASHTO Green Book gives:

$$\text{For } S > L, \; L = 2S - \frac{800}{A}\left(C - \frac{h_1 + h_2}{2}\right)$$

Given $C = 13.2$, $h_1 = 7.6$, $h_2 = 2.0$, $L = 1,496.54$ ft, $A = 9$, this equation yields $S = 1,121.6$ ft. This solution does not satisfy the criterion ($S > L$).

$$\text{For } S \le L, \; L = \frac{AS^2}{800\left(C - \dfrac{h_1 + h_2}{2}\right)}$$

Given $C = 13.2$, $h_1 = 7.6$, $h_2 = 2.0$, $L = 1,496.54$ ft, $A = 9$, this equation yields $S = 1,057$ ft. This solution satisfies the criterion ($S \le L$).

Corresponding design speed = 87.5 mph (impractical, of course, but solely based on sight distance under overhead obstruction). **D**

## 503

According to Shortt formula, the length of spiral:
$L_s = 1.6S^3/R = 1.6 \times 50^3/1,245 = 160$ ft **C**

## 504

Directional ADT $= 0.6 \times 65,000 = 39,000$ veh/day

Design hourly volume: DHV $= K \times$ ADT $= 0.12 \times 39,000 = 4,680$ veh/day

For 12-ft lanes, $f_{LW} = 0.0$

For 8-ft shoulders, $f_{LC} = 0.0$

For interchange density $= 1/1.25 = 0.8$ interchanges per mile, $f_{ID} = 1.5$ mph

For rural freeway segment, $f_N = 0.0$

For rural freeway, default value of BFFS = 75 mph

$$\text{FFS} = \text{BFFS} - f_N - f_{ID} - f_{LC} - f_{LW} = 75 - 0 - 1.5 - 0 - 0 = 73.5 \text{ mph}$$

$$f_{HV} = \frac{1}{1 + P_T(E_T - 1) + P_R(E_R - 1)} = \frac{1}{1 + 0.11 \times (2.5 - 1) + 0.05 \times (2.0 - 1)} = 0.82$$

$$v_p = \frac{V}{N f_P f_{HV} \text{PHF}} = \frac{4680}{3 \times 1.0 \times 0.82 \times 0.9} = 2,114$$

LOS D (range: 1,810 to 2,164 – values interpolated from HCM table) **D**

## 505

According to the Gravity model, the number of trips attracted from zone $i$ to zone $j$ is given in terms of trip production of zone $i$ ($P_i$) and trip attractions of zone $j$ ($A_j$) by:

$$T_{ij} = P_i\left[\frac{A_j F_{ij}}{\sum_j A_j F_{ij}}\right] = 440 \times \left[\frac{350 \times 35}{120 \times 75 + 350 \times 35 + 670 \times 40}\right] = 112 \qquad \text{A}$$

## 506

Number of trucks = $23,000 \times 0.12 = 2,760$ per day

Axle load data is based on 1,078 trucks. Therefore, results must be scaled by factor $2,760/1,078 = 2.56$

Growth percentage = 4%

Annual ADT numbers represent a geometric series (20 terms, $r = 1.04$). The cumulative ADT over 20 years is the sum of the series given by:

$$S = a[(r^n - 1)/(r - 1)] = ADT_1[(1.04^{20} - 1)/(1.04 - 1)] = 29.778 \times ADT_1$$

The year 1 *daily* ESAL: $W_{18} = \sum N_i LEF_i = 2,420 \times 0.0877 + 630 \times 0.36 + 301 \times 1 + 22 \times 1.51 + 6 \times 2.18 + 1 \times 3.53 + 24 \times 0.18 + 15 \times 0.308 + 12 \times 0.495 + 11 \times 0.857 = 814$

Cumulative 20-year ESAL: $W_{18} = 814 \times 365 \times 29.778 \times 2.56 = 22.65$ million $\qquad$ D

## 507

Lane width = 11-ft lanes. From HCM, $f_{LW} = 1.9$ mph

Total lateral clearance = $LC_L + LC_R = 4 + 6 = 10$ ft. From HCM, $f_{LC} = 0.4$ mph

For divided highway with median, HCM gives $f_M = 0.0$ mph

Access point density = $5,280/600 = 8.8$ access points per mile. From HCM, $f_A = 2.2$ mph

For rural/suburban multilane highway, BFFS = 60 mph, FFS = BFFS $- f_{LW} - f_{LC} - f_M - f_A =$ $60 - 1.9 - 0.4 - 0 - 2.2 = 55.5$ mph

The traffic stream includes 8% trucks, 3% buses, and 2% RVs. Heavy vehicle factor is:

$$f_{HV} = \frac{1}{1 + P_T(E_T - 1) + P_R(E_R - 1)} = \frac{1}{1 + 0.11 \times (1.5 - 1) + 0.02 \times (1.2 - 1)} = 0.944$$

Peak flow rate is calculated as:

$$v_p = \frac{V}{Nf_p f_{HV} \text{PHF}} = \frac{3,440}{3 \times 1.0 \times 0.944 \times 0.88} = 1,380$$

LOS C (range: 1,000 to 1,430) **B**

## 508

Radius of circular curve: $R_c = 5{,}729.578/D = 5{,}729.578/2.5 = 2{,}291.83$ ft

Deflection angle: $I = 56°35'48'' = 56.5967$

Spiral angle: $I_s = L_s D/200 = 230 \times 2.5/200 = 2.875 = 0.05018$ rad

$$x_c = L_s \left(1 - \frac{I_s^2}{10}\right) = 230 \times \left(1 - \frac{0.05018^2}{10}\right) = 229.94 \text{ ft}$$

$$y_c \approx \frac{L_s^2}{6R_c} = \frac{230^2}{6 \times 2{,}291.83} = 3.847 \text{ ft}$$

Coordinates of PC with respect to TS:

$$k = x_c - R_c \sin I_s = 229.94 - 2{,}291.83 \times \sin 2.875 = 114.99 \text{ ft}$$

$$p = y_c - R_c(1 - \cos I_s) = 3.847 - 2{,}291.83 \times (1 - \cos 2.875) = 0.9624 \text{ ft}$$

Spiral tangent length:

$$T_s = (R_c + p)\tan\frac{I}{2} + k = (2{,}291.83 + 0.9624) \times \tan\frac{56.5967°}{2} + 114.99 = 1{,}349.45 \text{ ft}$$

TS = (100 + 0.00) − (13 + 49.45) = 86 + 50.55 **A**

## 509

| Phase | Design Lane Group Volume | Saturation Volume for Design Lane Group |
|-------|--------------------------|------------------------------------------|
| 1 | 100 | 900 |
| 2 | 600 | 1,900 |
| 3 | 105 | 900 |
| 4 | 630 | 2,300 |

Total lost time = 12 sec

Webster's formula gives the optimum cycle length as:

$$C = \frac{1 \cdot 5 t_L + 5}{1 - \sum Y_i} = \frac{1 \cdot 5 \times 12 + 5}{1 - \left(\dfrac{100}{900} + \dfrac{600}{1,900} + \dfrac{105}{900} + \dfrac{630}{2,300}\right)} = \frac{23}{1 - 0.8175} = 126 \text{ sec}$$

B

## 510

$$f_{HV} = \frac{1}{1 + P_{HV}(E_{HV} - 1)} = \frac{1}{1 + 0.05 \times (2.0 - 1)} = 0.952$$

With 25 parking maneuvers/hr: $f_p = \dfrac{N - 0 \cdot 1 - \dfrac{18 N_m}{3,600}}{N} = \dfrac{2 - 0 \cdot 1 - \dfrac{18 \times 25}{3,600}}{2} = 0.888$

With 10 buses/hr: $f_{bb} = \dfrac{N - \dfrac{14.4 N_B}{3600}}{N} = \dfrac{2 - 14.4 \times \dfrac{10}{3600}}{2} = 0.98$

Saturation flow rate on WBTH/R approach (assuming non-CBD area):

$$s = s_0 N f_w f_{HV} f_g f_p f_{bb} f_a f_{RT} f_{LT} = 1,900 \times 2 \times 1.0 \times 0.952 \times 0.995 \times 0.888 \times 0.98 \times 1.0 \times$$
$$0.9625 \times 0.95$$
$$= 2864 \text{ pcph}$$

B

## 511

The clearance interval (for the wider pavement width = 60 ft) is calculated as:

$$\tau_{min} = t_R + \frac{V}{2a} + \frac{W + L}{V} = 1.0 + \frac{1.47 \times 40}{2 \times 10} + \frac{60 + 20}{1 \cdot 47 \times 40} = 5.3 \text{ sec}$$

C

## 512

Length of walkway = 60 ft

Cycle time = 80 sec; number of cycles per hour = 3,600/80 = 45

Number of pedestrians per cycle $N_{ped} = \dfrac{1,200}{45} = 26.67$

Green time: $G_p = 3.2 + \dfrac{L}{S_p} + 0.27 N_{ped} = 3.2 + \dfrac{60}{4.0} + 0.27 \times 26.67 = 25.4$ sec    A

## 513

Degree of curve = 4

Therefore, curve radius: $R = \dfrac{5,729.578}{D} = 1,432.4$ ft

Tangent length: $T = R \tan \dfrac{I}{2} = 768.25$ ft

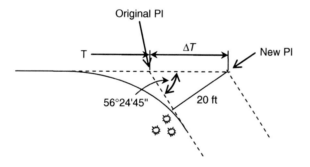

Original PI

$\Delta T$

T

New PI

56°24'45"    20 ft

From the triangle formed by the original PI, the new PI and the 20 ft offset, the tangent length will get longer by $\Delta T = \dfrac{20}{\sin 56°24'45''} = 24.01$ ft

New tangent length: $T + \Delta T = 768.25 + 24.01 = 792.26$ ft

New radius: $R' = T'/\tan \dfrac{I}{2} = 792.26/\tan 28.21 = 1,477.17$ ft

New degree of curve: $D' = 5,729.578/1,477.17 = 3.88$    C

## 514

Warrant 1:          Condition A – 600 major and 150 on minor

Condition B – 900 major and 75 on minor

Neither is met for *all* 8 hr at 100% level. Both are not met for all 8 hr at 80% level. Therefore, warrant 1 is not met.

Warrant 2: Is met for hours 4, 6, 7, and 8.

Warrant 3: Meets condition (2). Point plots above line for hr 7.

*Warrants 2 and 3 are met.*                                                                    D

## 515

All lateral and vertical force components are shown on the figure. A vertical plane is drawn through the heel of the footing.

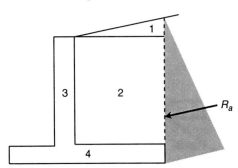

The height H, measured vertically from the bottom of the footing to the top of the backfill, $H = 17 + 6 \tan 15 = 18.6$ ft.

For $\phi = 34°$, $\beta = 15°$, $\delta = 0°$, $\theta = 90°$, Rankine's active earth pressure coefficient $K_a = 0.339$.

Earth pressure resultant: $R_a = \dfrac{1}{2}K_a \gamma H^2 = 0.5 \times 0.339 \times 120 \times 18.6^2 = 7{,}037\dfrac{\text{lb}}{\text{ft}}$

This can be resolved into its vertical and horizontal components as below:

$R_{av} = 7{,}037 \times \sin 15 = 1{,}821$ lb/ft

$R_{ah} = 7{,}037 \times \cos 15 = 6{,}797$ lb/ft

The weights of the soil wedges 1, 2, and the concrete elements 3 and 4 are calculated below, with their horizontal distances (measured from the toe).

$W_1 = \frac{1}{2} \times 6 \times 1.6 \times 120 = 576$ @ 9 ft

$W_2 = 14 \times 6 \times 120 = 10,080$ @ 8 ft

$W_3 = 1 \times 14 \times 150 = 2,100$ @ 4.5 ft

$W_4 = 11 \times 3 \times 150 = 4,950$ @ 5.5 ft

$R_{av} = 1,821$ @ 11 ft

The sum of the stabilizing moments about the toe is given by:

$$M_s = \sum F_i x_i = 576 \times 9 + 10,080 \times 8 + 2,100 \times 4.5 + 4,950 \times 5.5 + 1,821 \times 11$$
$$= 142,530 \text{ lb} \cdot \text{ft/ft}$$

The overturning moment about the toe is given by:

$$M_{OT} = R_{ah} \times H/3 = 6,797 \times 18.6/3 = 42,141 \text{ lb} \cdot \text{ft/ft}$$

$$FS_{OT} = 142,530/42,141 = 3.38 \hspace{6cm} \text{C}$$

## 516

Upstream peak flow rate: $V_F = 4,100 + 500 = 4,600$

Adjusted flow entering lanes 1 & 2 immediately upstream of diverge influence area is calculated as $V_{12} = V_R + (V_R - V_R)P_{FD}$.

For 6-lane freeways, there are three possible equations for $P_{FD}$. No upstream ramp and an adjacent downstream off-ramp, equation 7 or 5 applies. Calculate $L_{EQ}$.

$$L_{EQ} = \frac{V_D}{1.15 - 0.000032V_F - 0.000369V_R} = \frac{450}{1.15 - 0.000032 \times 4,600 - 0.000369 \times 500}$$
$$= 550 \text{ ft}$$

As the distance to adjacent downstream ramp is greater than $L_{EQ}$, use equation 5.

$$P_{FD} = 0.76 - 0.000025V_F - 0.000046V_R = 0.76 - 0.000025 \times 4,600 - 0.000046 \times 500$$
$$= 0.622$$

$$V_{12} = 500 + (4,600 - 500) \times 0.622 = 3,050 \text{ pcph} \hspace{4cm} \text{D}$$

### 517

In the matrix, zeros imply no link, "1" implies bidirectional link, "−1" implies restricted link (one way road). There are two one-way links in the network (4 to 2 and 5 to 6).

Counting all the nonzero entries adds up to 28. Divided by 2 (2 nodes per link), gives us the number of links, 14.　　　**A**

### 518

For ADT = 16,000, design speed = 65 mph and foreslope = 1:4, width of clear zone = 46 mph. This includes the shoulder width.

Therefore, clear distance from edge of shoulder = 46 − 8 = 38 ft　　　**D**

### 519

Segment capacity = 3,200 pc/h (for both directions) or 1,700 pc/h for each direction

Flow rate (both directions) = 2,800, does not exceed 3,200

Flow rate (design direction) = 0.6 × 2,800 = 1,680, does not exceed 1,700

From exhibit 20-2 of the Highway Capacity Manual, for average travel speed (ATS) = 42 mph, level of service is $D$; for percent time spent following (PTSF) = 63, level of service is C.

Governing LOS = $D$　　　**C**

### 520

Azimuth of back tangent = $180° + 42°30' = 222°30'$

Azimuth of forward tangent = $360° − 70° = 290°00'$

Deflection angle between tangents, I = $290°00' − 222°30' = 67°30'$

Radius: $R = 5{,}729.578/D = 5{,}729.578/3.75 = 1{,}527.89$ ft

Length of curve: $L = 100I/D = 100 \times 67.5/3.75 = 1{,}800$ ft

Tangent length: $T = R \tan \dfrac{I}{2} = 1{,}527.89 \times \tan \dfrac{67.5}{2} = 1{,}020.90$ ft

PI station: $50 + 22.30$

PC station $=$ PI station $- T = (50 + 22.30) - (10 + 20.90) = 40 + 01.40$

Arc length from PC to station $57 + 00.00 = 1{,}698.6$ ft

Deflection angle: $\alpha = \text{arc} \times \dfrac{I}{2L} = 1{,}698.6 \times \dfrac{67.5}{2 \times 1{,}800} = 31.849° = 31°50'56''$      A

## 521

Distance from PVC to PVI $= 6{,}009.0 - 5{,}312.50 = 696.50$ ft

Length of curve $= 2 \times 696.50 = 1{,}393$ ft $= 13.93$ stations

Rate of grade change: $R = (G_2 - G_1)/L = [3 - (-5)]/13.93 = 0.5743\%/\text{sta}$

Elevation of PVC: $y_{PVC} = (y_{PVI} - G_1)L/2 = 365.57 - (-5) \times 6.965 = 400.40$ ft

Bridge (at sta. $55 + 05.20$) is located at $x = 55.052 - 53.125 = 1.927$ stations

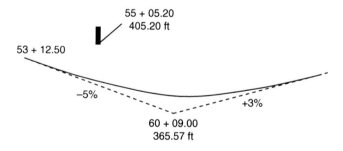

Curve elevation at location of bridge:

$$y = y_{PVC} + G_1 x + \tfrac{1}{2} R x^2 = 400.40 + (-5) \times 1.927 + \tfrac{1}{2} \times 0.5743 \times 1.927^2 = 391.83 \text{ ft}$$

Vertical clearance $= 405.20 - 391.83 = 13.37$ ft      B

## 522

Radius of roadway centerline = 750 ft

Radius of inside lane centerline = 750 − 6 = 744 ft

Middle ordinate distance (obstruction edge to centerline of inside lane) = 10 + 6 = 16 ft

$$S = \frac{R}{28.65}\cos^{-1}\left(1 - \frac{M}{R}\right) = \frac{744}{28.65}\cos^{-1}\left(1 - \frac{16}{744}\right) = 309.13 \text{ ft}$$

From table for stopping sight distance (AASHTO Green Book), design speed for stopping sight distance is 309 ft (level roadway), which is just above 40 mph.　　**B**

## 523

According to the 2009 MUTCD, the minimum warning distance is 175 ft (Table 2C-4).　　**D**

### Table 2C-4. Guidelines for Advance Placement of Warning Signs

| Posted or 85th-Percentile Speed | Advance Placement Distance[1] | | | | | | | | |
|---|---|---|---|---|---|---|---|---|---|
| | Condition A: Speed reduction and lane changing in heavy traffic[c] | Condition B: Deceleration to the listed advisory speed (mph) for the condition | | | | | | | |
| | | 0[3] | 10[4] | 20[4] | 30[4] | 40[4] | 50[4] | 60[4] | 70[4] |
| 20 mph | 225 ft | 100 ft[6] | N/A[5] | — | — | — | — | — | — |
| 25 mph | 325 ft | 100 ft[6] | N/A[5] | N/A[5] | — | — | — | — | — |
| 30 mph | 460 ft | 100 ft[6] | N/A[5] | N/A[5] | — | — | — | — | — |
| 35 mph | 565 ft | 100 ft[6] | N/A[5] | N/A[5] | N/A[6] | — | — | — | — |
| 40 mph | 670 ft | 125 ft | 100 ft[6] | 100 ft[6] | N/A[5] | — | — | — | — |
| 45 mph | 775 ft | 175 ft | 125 ft | 100 ft[6] | 100 ft[6] | N/A[5] | — | — | — |
| 50 mph | 885 ft | 250 ft | 200 ft | 175 ft | 125 ft | 100 ft[6] | — | — | — |
| 55 mph | 990 ft | 325 ft | 275 ft | 225 ft | 200 ft | 125 ft | N/A[5] | — | — |
| 60 mph | 1,100 ft | 400 ft | 350 ft | 325 ft | 275 ft | 200 ft | 100 ft[6] | — | — |
| 65 mph | 1,200 ft | 475 ft | 450 ft | 400 ft | 350 ft | 275 ft | 200 ft | 100 ft[6] | — |
| 70 mph | 1,250 ft | 550 ft | 525 ft | 500 ft | 450 ft | 375 ft | 275 ft | 150 ft | — |
| 75 mph | 1,350 ft | 650 ft | 625 ft | 600 ft | 550 ft | 475 ft | 375 ft | 250 ft | 100 ft[6] |

## 524

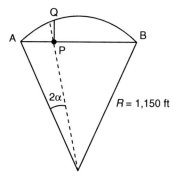

Length along chord:

$$2R\sin\alpha\cos\left(\frac{I}{2}-\alpha\right)=215\Rightarrow\sin\alpha\cos(18-\alpha)=0.0935\Rightarrow\alpha$$

$$=5.5°\text{ (by trial and error)}$$

The length of the offset from the chord (PQ) is given by:

$$2R\sin\alpha\sin\left(\frac{I}{2}-\alpha\right)=2\times1,150\times\sin5.5\times\sin(18-5.5)=47.71\text{ ft}$$

B

## 525

$$A=\left|G_2-G_1\right|=3.6\%$$

Design speed = 65 mph

Corresponding stopping sight distance (AASHTO Green Book) = reaction time distance (239 ft) + braking distance (405 ft) = 644 ft

Assuming $S<L$: $L=AS^2/2,158=(3.6\times644^2)/2,158=692$ ft

This solution fits the criterion $S<L$. Therefore, the solution is OK.

Minimum length of curve = 692 ft

D

## 526

Actual space-hour utilization is calculated as the sum $\Sigma N_i h_i$, where $N_i$ is the number of cars in a particular group and $h_i$ is the number of hours these cars are parked.

$$\Sigma N_i h_i = 360 \times (0.1 \times 1 + 0.15 \times 2 + 0.2 \times 3 + 0.3 \times 4 + 0.25 \times 10) = 1{,}692 \text{ space} \cdot \text{hr}$$

This represents 85% occupancy. Therefore, space hours available is $1{,}692/0.85 = 1{,}990$.

Assuming $N$ spaces in the lot, with the lot operating 10 hr a day, with 80% efficiency, number of space-hours available $= 0.8 \times 10 \times N = 8N$, which must be equal to 1,990. Thus $N = 248.8$.                     **A**

## 527

Design hourly volume, $\text{DHV} = K \times \text{ADT} = 0.12 \times 52{,}500 = 6{,}300 \text{ veh/hr}$

Directional design hourly volume, $\text{DDHV} = D \times \text{DHV} = 0.6 \times 6{,}300 = 3{,}780 \text{ veh/hr}$

Peak hour factor, $\text{PHF} = 0.9$

Number of lanes in each direction, $N = 3$

$\text{FFS} = 70$

$$f_{HV} = \frac{1}{1 + P_T(E_T - 1) + P_R(E_R - 1)} = \frac{1}{1 + 0.11 \times (4.5 - 1) + 0.05 \times (4.0 - 1)} = 0.651$$

Peak flow rate, $v_p = V/Nf_p f_{HV} \text{PHF} = 3{,}780/(3 \times 1.0 \times 0.651 \times 0.9) = 2{,}151 \text{ pcphpl}$

For an $\text{FFS} = 70$ mph, peak flow rate $v_p = 2{,}151$ yields LOS E                     **D**

## 528

Since the new highway is parallel to the original, the two curves are identical (same radius, same deflection angle).

Radius: $R = 5{,}729.578/D = 5{,}729.578/2.0 = 2{,}864.79$ ft

Length of major chord: $L_{chord} = 2R \sin\dfrac{I}{2}$

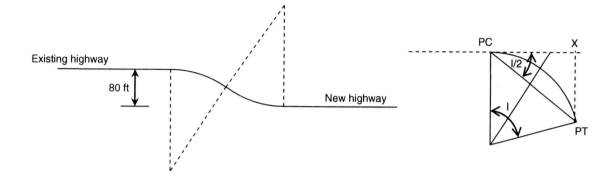

X is the point where the tangent offset passes through the PT. In this problem, this offset is 40 ft (half of 80 ft). Note that if $I \neq 90°$, point X is not the same as the PI. The trigonometry of this triangle (PC-PT-X) yields:

$$\frac{2R \sin^2 I}{2} = 40 \Rightarrow I = 9.586°$$

Length of each curve: $L = \frac{100I}{D} = \frac{100 \times 9.586}{2} = 479.29$ ft

Total length of reverse curve = $2 \times 479.29 = 958.58$ ft          C

## 529

Highway Capacity Manual 2000, Chapter 17.

From exhibit 17-5, for left turn from minor street, where the major street has four lanes, base critical gap, $t_{c,base} = 7.5$ sec

Adjustment factor for heavy vehicles, $t_{c,HV} = 2.0$ for 4-lane major street

Adjustment factor for grade, $t_{c,G} = 0.2$ for left turn from minor street

Adjustment factor for two-stage gap acceptance, $t_{c,T} = 0$ (since this is single stage)

Adjustment factor for intersection geometry (0.7 for minor street left turn movement)

$$t_{c,X} = t_{c,base} + t_{c,HV}P_{HV} + t_{c,G}G - t_{c,T} - t_{3,LT} = 7.5 + 2.0 \times 0.15 + 0.2 \times 0.04 - 0 - 0.7$$

$$= 7.1 \text{ sec} \qquad\qquad \text{A}$$

## 530

Jam density $D_j = 64$ veh/mi

Free flow speed $S_f =$ twice the optimum speed $= 100$ mph

Capacity, $c = \dfrac{1}{4} S_f D_j = \dfrac{1}{4} \times 100 \times 64 = 1{,}600$ veh/hr     **B**

*Note:* Greenshield's (linear) model has been used here.

$$S = S_f \left(1 - \dfrac{D}{D_j}\right)$$

## 531

$F_{200} = 8$, LL $= 43$, PI $= 43 - 21 = 22$

According to the chart below (from I. Goswami, *Civil Engineering PE All-in-One Exam Guide: Breadth and Depth*, McGraw-Hill, 2009), the criteria $F_{200} < 35$, LL $> 40$ & PI $> 10$ matches group A-2-7.     **D**

**Table 203.5 AASHTO Soil Classification Criteria**

| Sieve Analysis | A-1-a | A-1-b | A-3 | A-2-4 | A-2-5 | A-2-6 | A-2-7 | A-4 | A-5 | A-6 | A-7 | A-8 |
|---|---|---|---|---|---|---|---|---|---|---|---|---|
| % passing | | | | | | | | | | | | |
| No. 10 | ≤50 | | | | | | | | | | | |
| No. 40 | ≤30 | ≤50 | >50 | | | | | | | | | |
| No. 200 | ≤15 | ≤25 | ≤10 | ≤35 | ≤35 | ≤35 | ≤35 | >35 | >35 | >35 | >35 | |
| LL | | | | ≤40 | >40 | ≤40 | >40 | ≤40 | >40 | ≤40 | >40 | |
| PI | ≤6 | | NP | ≤10 | ≤10 | >10 | >10 | ≤10 | ≤10 | >10 | >10 | |

Granular Materials (35% or less passing no. 200 sieve) covers A-1 (A-1-a, A-1-b), A-3, and A-2 (A-2-4, A-2-5, A-2-6, A-2-7). Silt-Clay Materials (more than 35% passing no. 200 sieve) covers A-4, A-5, A-6, A-7.

## 532

After the base-flow is subtracted, we get:

| Time (hr) | 0 | 1 | 2 | 3 | 4 | 5 | 6 |
|---|---|---|---|---|---|---|---|
| Discharge Q (ft³/sec) | 0 | 61 | 104 | 89 | 52 | 9 | 2 |

Volume of runoff: $V = \Sigma Q \times \Delta t = 317 \times 3{,}600 = 1{,}141{,}200 \text{ ft}^3$

Average depth of runoff (excess precipitation) = $1{,}141{,}200/(115 \times 43{,}560)$ = 0.228 ft = 2.73 in.

The peak *runoff* discharge (ft³/sec) that would be recorded following a 2-hr storm producing 1.7 in. of runoff is most nearly $104 \times 1.7/2.73 = 65 \text{ ft}^3/\text{sec}$.

Adding in the base flow (unaffected by storm), the peak discharge that would be recorded in the stream is 88 ft³/sec. **A**

## 533

For pipe diameter = 36 in. and $Q = 100$ cfs, FHWA nomograph yields $HW/D = 3.5$

Headwater = $3.5 \times 3 = 10.5$ ft (this is measured from the upstream invert elevation)

100-year water surface elevation = $325 + 10.5 = 335.5$ ft

Vertical clearance between roadway and 100-year WSE = $337.8 - 335.5 = 2.3$ ft **B**

## 534

This is an "unknown-depth-of-flow" problem. It is best solved using the tables (see Table 303.3 and Equation 303.37 in *All-in-One*).

Flow parameter: $K = \dfrac{Qn}{kb^{8/3}S^{1/2}} = \dfrac{150 \times 0.016}{1.486 \times 10^{8/3} \times 0.004^{1/2}} = 0.055$

*(Continued on next page)*

For the m = 2 (side slope parameter), the fragment of the table is shown below:

| d/b | 0 | 0.25 | 0.5 | 0.75 | 1.0 | 1.5 | 2.0 | 2.5 | 3.0 | 3.5 | 4.0 |
|---|---|---|---|---|---|---|---|---|---|---|---|
| | | | | | | | Horizontal projection m | | | | |
| 0.01 | 0.0005 | 0.0005 | 0.0005 | 0.0005 | 0.0005 | 0.0005 | 0.0005 | 0.0005 | 0.0005 | 0.0005 | 0.0005 |
| 0.02 | 0.0014 | 0.0014 | 0.0015 | 0.0015 | 0.0015 | 0.0015 | 0.0015 | 0.0015 | 0.0015 | 0.0015 | 0.0015 |
| 0.04 | 0.0044 | 0.0045 | 0.0046 | 0.0046 | 0.0047 | 0.0047 | 0.0048 | 0.0048 | 0.0049 | 0.0049 | 0.0050 |
| 0.06 | 0.0085 | 0.0087 | 0.0089 | 0.0090 | 0.0091 | 0.0093 | 0.0095 | 0.0096 | 0.0098 | 0.0099 | 0.0101 |
| 0.08 | 0.0135 | 0.0139 | 0.0142 | 0.0145 | 0.0147 | 0.0152 | 0.0155 | 0.0159 | 0.0162 | 0.0165 | 0.0168 |
| 0.1 | 0.0191 | 0.0198 | 0.0204 | 0.0209 | 0.0214 | 0.0221 | 0.0228 | 0.0234 | 0.0241 | 0.0247 | 0.0253 |
| 0.12 | 0.0253 | 0.0265 | 0.0275 | 0.0283 | 0.0290 | 0.0303 | 0.0314 | 0.0324 | 0.0334 | 0.0345 | 0.0355 |
| 0.14 | 0.0320 | 0.0338 | 0.0352 | 0.0365 | 0.0376 | 0.0395 | 0.0412 | 0.0428 | 0.0444 | 0.0459 | 0.0475 |
| 0.16 | 0.0392 | 0.0416 | 0.0437 | 0.0455 | 0.0471 | 0.0498 | 0.0523 | 0.0546 | 0.0569 | 0.0591 | 0.0614 |
| 0.18 | 0.0467 | 0.0500 | 0.0529 | 0.0553 | 0.0575 | 0.0612 | 0.0646 | 0.0678 | 0.0710 | 0.0741 | 0.0772 |
| 0.2 | 0.0547 | 0.0589 | 0.0627 | 0.0659 | 0.0687 | 0.0737 | 0.0783 | 0.0826 | 0.0868 | 0.0910 | 0.0952 |
| 0.22 | 0.0629 | 0.0683 | 0.0731 | 0.0772 | 0.0809 | 0.0874 | 0.0932 | 0.0989 | 0.1043 | 0.1098 | 0.1152 |
| 0.24 | 0.0714 | 0.0781 | 0.0841 | 0.0893 | 0.0939 | 0.1021 | 0.1096 | 0.1167 | 0.1237 | 0.1306 | 0.1374 |

Interpolating, $d/b = 0.164$

$d = 1.64$ ft = 19.7 in.

C

# 535

By examination, it seems that the peak dry unit weight will come from sample 3 or 4.

Volume of Standard Proctor mold = $1/30$ ft$^3$

For sample 3, total unit weight, $\gamma = W/V = 3.95/(1/30) = 118.5$ lb/ft$^3$

Dry unit weight, $\gamma_d = 118.5/1.16 = 102.2$ lb/ft$^3$

For sample 4, total unit weight, $\gamma = W/V = 4.21/(1/30) = 126.3$ lb/ft$^3$

Dry unit weight, $\gamma_d = 126.3/1.18 = 107.0$ lb/ft$^3$ (maximum dry unit weight)

Target dry unit weight = $0.9 \times 107 = 96.3$ lb/ft$^3$

Weight of solids needed = $96.3 \times 1.5 \times 10^6 \times 27 = 3.9 \times 10^9$ lb

From table, for 12% moisture content, unit weight = 3.24/(1/30) = 97.2 lb/ft³; $\gamma_d$ = 97.2/1.12 = 86.8 lb/ft³.

Volume of borrow soil needed = 3.9 × 10⁹ lb/86.8 lb/ft³ = 4.49 × 10⁷ ft³ = 1.66 × 10⁶ yd³   **B**

## 536

Number of entering vehicles (24 hr) = 5,720 (ADT)

Accident rate:

$R$ = No. of accidents × 10⁸/ADT × 365 × $N$ = 15 × 10⁸/5,720 × 365 × 1 = 718

Accident rate = 718 accidents per hundred million entering vehicles   **C**

## 537

According to the 2009 MUTCD, the location of the ROAD WORK AHEAD sign from the center of the work zone = 1,250/2 + distances A, B, and C = 625 + 1,000 + 1,500 + 2,640 = 5,765 ft (see table below).   **B**

### Table 6H-3. Meaning of Letter Codes on Typical Application Diagrams

| Road Type | Distance Between Signs** | | |
|---|---|---|---|
| | A | B | C |
| Urban (low speed)* | 100 feet | 100 feet | 100 feet |
| Urban (high speed)* | 350 feet | 350 feet | 350 feet |
| Rural | 500 feet | 500 feet | 500 feet |
| Expressway / Freeway | 1,000 feet | 1,500 feet | 2,640 feet |

## 538

MUTCD figure 6F-1 shows vertical clearance under signs in an urban environment = 7 ft   **C**

## 539

Using average end area method, and adjusting the fill volumes by 1.12 factor, we get the following:

| Station | Area (ft²) | | Volume (ft³) | | | |
|---|---|---|---|---|---|---|
| | Cut | Fill | Cut | Adjusted Fill | Net | Cumulative |
| 0 + 00.00 | 563.2 | 342.2 | | | | |
| 0 + 50.00 | 213.5 | 213.6 | 719.2 | −576.4 | 142.8 | 142.8 |
| 1 + 00.00 | 123.5 | 343.3 | 312.0 | −577.5 | −265.5 | −122.7 |
| 1 + 50.00 | 654.6 | 111.0 | 720.5 | −471.1 | 249.3 | 126.6 |
| 2 + 00.00 | 973.1 | 762.4 | 1,507.1 | −905.7 | 601.4 | 728.0 |
| 2 + 50.00 | 567.3 | 342.9 | 1,426.3 | −1,146.2 | 280.0 | 1,008.0 |
| 3 + 00.00 | 451.6 | 190.4 | 943.4 | −553.1 | 390.4 | 1,398.4 |

Net earthwork volume = 1,398.4 cu yd                                       A

*Note:* Even though shown in the table above, the last two columns do not need to be constructed. Sum of all cuts = 5,628.5. Sum of all adjusted fills = 4,230.1. Desired answer = 5,628.5 − 4,230.1 = 1,398.4. 540.

## 540

For a perpetual annuity, capitalized cost = $A/i$

All costs are calculated in millions of dollars.

Present worth of Plan A (single capital investment) is calculated as the sum of initial expense + present worth of perpetual annuity.

$$P_A = 4.2 + 0.04 \times \frac{1}{0.07} = 4.771$$

Present worth of Plan B (two-stage capital investment) is calculated as the sum of initial expense + lump-sum expense at $t = 10$ years + perpetual annuity of $16 + 10$ year annuity of 4.

$$P_B = 2.0 + 3.2\left(\frac{P}{F}, 10 \text{ years}, 7\%\right) + 0.036 \times \frac{1}{0.07} - 0.008\left(\frac{P}{A}, 10 \text{ years}, 7\%\right)$$

$$= 2.0 + 3.2 \times 0.5083 + 0.036 \div 0.07 - 0.008 \times 7.0236 = 4.085$$

Cost ratio = 4.085/4.771 = 0.856                                       A

## Answer Key for Transportation Depth Exam

| | | | | | | | | |
|---|---|---|---|---|---|---|---|
| 501 | C | | 511 | C | | 521 | B |
| 502 | D | | 512 | A | | 522 | B |
| 503 | C | | 513 | C | | 523 | D |
| 504 | D | | 514 | D | | 524 | B |
| 505 | A | | 515 | C | | 525 | D |
| 506 | D | | 516 | D | | 526 | A |
| 507 | B | | 517 | A | | 527 | D |
| 508 | A | | 518 | D | | 528 | C |
| 509 | B | | 519 | C | | 529 | A |
| 510 | B | | 520 | A | | 530 | B |

| | |
|---|---|
| 531 | D |
| 532 | A |
| 533 | B |
| 534 | C |
| 535 | B |
| 536 | C |
| 537 | B |
| 538 | C |
| 539 | A |
| 540 | A |

# 14

# Construction Depth Exam Solutions

These detailed solutions are for questions 601 to 640, representative of a 4-hour construction depth exam according to the syllabus and guidelines for the Principles and Practice (P&P) of Civil Engineering Examination, administered by the National Council of Examiners for Engineering and Surveying (NCEES).

## 601

Cycle time of the excavator = 45 sec + 4 min + 30 sec = 5 min 15 sec = 5.25 min

Ideal volume (loose measure) turnover = 2.8 yd$^3$/5.25 min = 32 yd$^3$/hr

Actual volume turnover (8-hr day) = 0.85 × 32 = 27.2 yd$^3$/hr = 217.6 yd$^3$/day

Excavated material = 50,000 ft$^3$ (bank measure) = 1.2 × 50,000 ft$^3$ = 60,000 ft$^3$ (loose) = 2,222 yd$^3$

The number of days required to complete the job = 2,222/217.6 = 10.2 days

Round up to 11 days.                                                                              C

## 602

Cycle time of the excavator = 45 sec + 4 min + 45 sec = 5 min 30 sec = 5.5 min

Ideal volume (loose measure) turnover = 3.0 yd³/5.5 min = 0.545 yd³/min

Actual volume turnover = 0.9 × 0.545 = 0.491 yd³/min

Truck production = 15 yd³/90 min = 0.167 yd³/min/truck

The number of trucks needed to balance the production of the excavator = 0.491/0.167 = 2.95

Use 3 trucks.

**B**

*Note:* Since this is a question of balance, the total size of the job does not matter.

## 603

Area of walls (single layer): A = 2×(180+200)×12−8×6×10 = 8,640 ft²

Fully burdened cost for labor:

4×40+2×25+1×50 = $260 per crew hr = $37.14 per labor hr

Credit for 5/8 in. GWB (1 layer):   Labor = 8,640 ft²/ 960 ft²/LH × $37.14/LH
= $334.27
Materials = 8,640 ft² × $0.255/ft² = $2,203.20 + 10% waste = $2,423.52
TOTAL = $2,757.79

½ in. GWB (2 layers):   Labor = 17,280 ft²/ 960 ft²/LH × $37.14/LH
= $668.52
Materials = 17,280 ft² × $0.285/ft² = $4,924.80 + 10% waste = $5,417.28
TOTAL = $6,085.80

Insulation (1 layer):   Labor = 8,640 ft²/1,920 ft²/LH × $37.14/LH
= $167.13
Materials = 8,640 ft² × $0.45/ft² = $3,888.00 + 10% waste = $4,276.80
TOTAL = $4,443.93

Cost differential (new SOW − original SOW) = −2,757.79 + $6,085.80 + 4,443.93 = $7,771.94

After applying 1.10 and 1.05 factors for overhead and profit, the net increase in the change order is $9,831.50.

C

## 604

As the side-slopes are 3:2 and the horizontal projection of the side-flares are 30 ft, the depth of the channel is 20 ft and the inclined sides are $\sqrt{30^2 + 20^2} = 36.1$ ft.

Total channel perimeter = 19 ft + 2 × 36.1 = 91.2 ft

Total surface area = 91.2 × 450 = 41,040 ft$^2$

Cost of finishing compound (including waste) = 41,040 ft$^2$/300 ft$^2$/gal × 1.05 = 143.64 gal = 28.7 × 5-gal containers. Use 29 containers. Cost = 29 × $40/container = $1,160.00.

Total cross sectional area = 91.2 ft × 8 in. = 60.8 ft$^2$

Total volume of concrete = 60.8 × 450 = 27,360 ft$^3$ = 1,013.33 yd$^3$

Cost of concrete (including waste) = 98 × 1,013.33 × 1.05 = $104,272

Total cost of materials = $1,160 + $104,272 = $105,432

D

## 605

Dump truck:    Ideal travel time = 8 mi round trip/30 mph = 0.267 hr = 960 sec

Ideal transfer time = 40 sec per 3 yd$^3$ = 213 sec per 16 yd$^3$ load

Ideal dumping time = 30 sec per load

Cycle time = 960 + 213 + 30 = 1,203 sec

Dump truck productivity (ideal) = 16 yd$^3$ per 1,203 sec = 0.0133 yd$^3$/sec (loose soil)

Dump truck productivity (actual) = 0.7 × 0.8 × 0.0133 = 0.00745 yd$^3$/sec (loose soil)

Rate of power shovel excavation = 10 yd$^3$/min (bank) = 11 yd$^3$/min (loose).

For each 3 yd$^3$ (loose soil) load of the power shovel, excavation time = 3/11 min = 16.4 sec

Cycle time of power shovel = excavation time + transfer time + dump time = 16.4 + 40 + 30 = 86.4 sec

(Continued on next page)

Power shovel production (loose soil) = 3 yd³/86.4 sec = 0.03472 yd³/sec

Number of dump trucks needed to balance the power shovel production =
0.03472/0.00745 = 4.66

Use 5 trucks.                                                                                    C

## 606

The network diagram constructed from the information in the table is shown below.
Each activity is shown as a block that shows [starting with upper left corner, going
clockwise–ES (early start), EF (early finish), LF (late finish), LS (late start)].

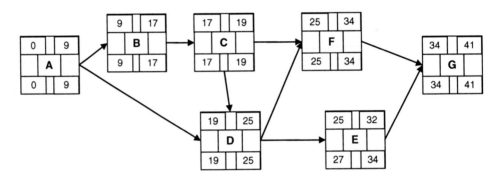

Project duration is the finish time for the terminal activity = 41 days                         C

## 607

Maximum moment, $M = \dfrac{PL}{3} = \dfrac{800 \times 60}{3} = 16{,}000 \text{ lb} \cdot \text{in.}$

Section modulus of rectangular beam section, $S = \dfrac{1}{6}bh^2 = \dfrac{6^3}{6} = 36 \text{ in.}^3$

Maximum bending stress, $\sigma = \dfrac{M}{S} = \dfrac{16{,}000}{36} = 444.44 \text{ psi}$

This is the modulus of rupture, which is correlated to $f_c'$ according to: $f_r = 7.5\sqrt{f_c'}$

Therefore, $f_c' = 3{,}511 \text{ psi}$                                                         D

## 608

Split cylinder test tensile strength: $f_{ct} = \dfrac{2P}{\pi DL} = \dfrac{2 \times 55,000}{\pi \times 6 \times 12} = 486$ psi

$f_{ct} = 6.7\sqrt{f_c'}$

Therefore, $f_c' = 5,262$ psi     **B**

## 609

For a pour rate of 12 ft/hr and concrete temperature of 50°F, the base value of lateral pressure in a column form is 2,310 psf.

For normal weight concrete, there is no adjustment. For cement containing less than 40% fly ash and no retarder, adjustment factor = 1.2.

Therefore, lateral pressure = 1.2 × 2,310 = 2,772 psf. However, the lateral pressure cannot exceed the hydrostatic pressure = 145 × 18 = 2,610 psf. Design pressure will be hydrostatic (linear) growing from zero at the top of the form to a maximum value = 2,610 psf a depth of 18 ft and stay constant at 2,610 psf.     **A**

## 610

According to MUTCD Table 6H-3, the location of the ROAD WORK AHEAD sign from the center of the work zone = 1,250/2 + distances A, B, and C = 625 + 1,000 + 1,500 + 2,640 = 5,765 ft     **B**

### Table 6H-3. Meaning of Letter Codes on Typical Application Diagrams

| Road Type | Distance Between Signs** | | |
|---|---|---|---|
| | A | B | C |
| Urban (low speed)* | 100 feet | 100 feet | 100 feet |
| Urban (high speed)* | 350 feet | 350 feet | 350 feet |
| Rural | 500 feet | 500 feet | 500 feet |
| Expressway / Freeway | 1,000 feet | 1,500 feet | 2,640 feet |

## 611

According to the *Standard Practice for Bracing Masonry Walls during Construction*, whenever a masonry wall is constructed, a limited access zone is to be established, whose width shall be wall height plus 4 ft.     **C**

## 612

According to OSHA 3146: "Anchorages used to attach personal fall arrest systems shall be independent of any anchorage being used to support or suspend platforms and must be capable of supporting at least 5,000 lb (22.2 kN) per person attached."

A

## 613

By examination, it seems that the peak dry unit weight will come from sample 3 or 4.

Volume of Standard Proctor mold = $1/30$ ft$^3$

For sample 3, total unit weight, $\gamma = W/V = 3.95/(1/30) = 118.5$ lb/ft$^3$

Dry unit weight, $\gamma_d = 118.5/1.16 = 102.2$ lb/ft$^3$

For sample 4, total unit weight, $\gamma = W/V = 4.21/(1/30) = 126.3$ lb/ft$^3$

Dry unit weight, $\gamma_d = 126.3/1.18 = 107.0$ lb/ft$^3$ (maximum dry unit weight)

Target dry unit weight, $\gamma_{d,max} = 0.9 \times 107 = 96.3$ lb/ft$^3$

Weight of solids needed = $96.3 \times 1.5 \times 10^6 \times 27 = 3.9 \times 10^9$ lb

From table, for 12% moisture content, unit weight = $3.24/(1/30) = 97.2$ lb/ft$^3$

$\gamma_d = 97.2/1.12 = 86.8$ lb/ft$^3$

Volume of borrow soil needed = $3.9 \times 10^9$ lb$/86.8$ lb/ft$^3 = 4.49 \times 10^7$ ft$^3$ = $1.66 \times 10^6$ yd$^3$

B

## 614

According to CFR1910 Table G-16, maximum impulsive noise exposure is 140 dB.

D

TABLE G–16—PERMISSIBLE NOISE EXPOSURES [1]

| Duration per day, hours | Sound level dBA slow response |
|---|---|
| 8 | 90 |
| 6 | 92 |
| 4 | 95 |
| 3 | 97 |
| 2 | 100 |
| 1½ | 102 |
| 1 | 105 |
| ½ | 110 |
| ¼ or less | 115 |

[1] When the daily noise exposure is composed of two or more periods of noise exposure of different levels, their combined effect should be considered, rather than the individual effect of each. If the sum of the following fractions: $C_1/T_1 + C_2/T_2 \ldots C_n/T_n$ exceeds unity, then the mixed exposure should be considered to exceed the limit value. $C_n$ indicates the total time of exposure at a specified noise level, and $T_n$ indicates the total time of exposure permitted at that level.

Exposure to impulsive or impact noise should not exceed 140 dB peak sound pressure level.

## 615

According to the Arrhenius model, considering the entire curing period (3 days) to have average temperature (28°C = 31 K), the equivalent age is calculated as:

$$t_{20} = e^{-\frac{41,500}{8.3144}\left(\frac{1}{301} - \frac{1}{293}\right)} \times 3 = 4.72$$

D

## 616

Volume of fill = 15,000 yd³ = 405,000 ft³

At a compacted unit weight of 140 pcf, this fill weighs = $5.67 \times 10^7$ lb = 28,350 tons = 1134 truckloads (each truckload = 50,000 lb net).

Roundtrip distance = 64 mi, which requires 64/32 = 2 hr travel + 35 min for loading and unloading. Therefore, cycle time for truck = 2.583 hr.

(*Continued on next page*)

Total hours truck is needed to complete the job = $1,134 \times 2.583 = 2,929$ hr

Truck operational cost = $\$80 \times 2,929 = \$234,330$

Fill material cost = $\$3/\text{ton} \times 28,350 = \$85,050$

Cost to place and compact fill = $15,000 \times 0.5 = \$7,500$

Total cost = $\$326,880$

Cost per yd$^3$ = $\$326,880/15,000 = \$21.79$                                                    D

## 617

Total number of hours worked by all employees = $400 \times 1,970 = 788,000$

Total recordable cases = $13 + 18 = 31$

OSHA normalizes this rate to 200,000 work hours.

OSHA recordable case rate = $31 \times 200,000/788,000 = 7.87$                          B

## 618

Using average end area method, and adjusting the fill volumes for shrinkage, we get the following:

| Station | Area (ft²) | | Volume (yd³) | | | | |
| | Cut | Fill | Cut | Fill | Adjusted Fill | Net | Cumulative |
|---|---|---|---|---|---|---|---|
| 0 + 00.00 | 563.2 | 342.2 | | | | | |
| 0 + 50.00 | 213.5 | 213.6 | 719.2 | −514.6 | −584.8 | 134.4 | 134.4 |
| 1 + 00.00 | 123.5 | 343.3 | 312.0 | −515.6 | −586.0 | −273.9 | −139.6 |
| 1 + 50.00 | 654.6 | 111.0 | 720.5 | −420.6 | −478.0 | 242.5 | 102.9 |
| 2 + 00.00 | 973.1 | 762.4 | 1,507.1 | −808/7 | −919.0 | 588.1 | 691.0 |
| 2 + 50.00 | 567.3 | 342.9 | 1,426.3 | −1,023.4 | −1,163.0 | 263.3 | 954.3 |
| 3 + 00.00 | 451.6 | 190.4 | 943.4 | −493.8 | −561.1 | 382.3 | 1,336.6 |

As an example: between station 0 + 0.00 and 0 + 50.00, volume of fill = $\frac{1}{2}(342.2 + 213.6) \times 50 = 13,895$ ft$^3$ = 514.6 yd$^3$

Adjusted fill = $514.6/0.88 = 584.8$ yd$^3$ (shown highlighted)

Net earthwork volume (sum of all adjusted fill and all cut volumes) = 1,336.6 yd³    A

*Note:* Even though shown in the table above, the last two columns do not need to be constructed. Sum of all cuts = 5,628.5. Sum of all adjusted fills = 4,291.9. Desired answer = 5,628.5 − 4,291.9 = 1,336.6.

## 619

Pursuing option B has:     Initial cost = $200,000

Monthly savings = $17,000 − 8,000 = $9,000

Resale value (after 18 months) = $120,000

Monthly interest rate (nominal) = 10/12 = 0.833%

$$\left(\frac{A}{F}, 0.833\%, 18 \text{ periods}\right) = \frac{i}{(1+i)^n - 1} = \frac{0.00833}{1.00833^{18} - 1} = 0.057124$$

$$\left(\frac{A}{P}, 0.833\%, 18 \text{ periods}\right) = \frac{i(1+i)^n}{(1+i)^n - 1} = \frac{0.00833 \times 1.00833^{18}}{1.00833^{18} - 1} = 0.060057$$

Converting everything to annuities:     Initial cost = 200,000 × 0.060057 = $12,011

Monthly benefit = 9,000

Cost offset (resale) = 120,000 × 0.051724 = $6,207

Benefit cost ratio: $\dfrac{B}{C} = \dfrac{B}{C-S} = \dfrac{9,000}{12,011 - 6,207} = 1.55$     C

*Note:* It is standard practice to designate the salvage value as a negative cost rather than a benefit, even though it represents "inward" cash flow.

## 620

It helps to recognize that in the given geometry, the lines AF and CD are parallel and so are AC and FD, and that these lines are of the same length, thereby forming a square.

(*Continued on next page*)

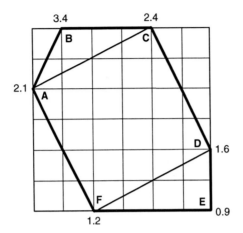

Thus, the volume can be found by adding together the volume for the shapes ABC, ACDF, and DEF.

Area of triangle ABC (base BC = 150 ft, height = 100 ft) is:

$$A_{ABC} = \frac{1}{2}bh = \frac{1}{2} \times 150 \times 100 = 7{,}500 \text{ ft}^2$$

Average depth of cut, $d_{ave} = \frac{1}{3}(2.1 + 3.4 + 2.4) = 2.63$ ft

Volume of cut in triangle ABC: $V_{ABC} = 7{,}500 \times 2.63 = 19{,}725 \text{ ft}^3$

Area of square ACDF (sides 223.6 ft) is:

$$A_{ACDF} = 223.6 \times 223.6 = 50{,}000 \text{ ft}^2$$

Average depth of cut, $d_{ave} = \frac{1}{4}(2.1 + 2.4 + 1.6 + 1.2) = 1.83$ ft

Volume of cut in square ACDF: $V_{ACDF} = 50{,}000 \times 1.83 = 91{,}500 \text{ ft}^3$

Area of triangle DEF (base EF = 200 ft, height DE = 100 ft) is:

$$A_{DEF} = \frac{1}{2}bh = \frac{1}{2} \times 200 \times 100 = 10{,}000 \text{ ft}^2$$

Average depth of cut, $d_{ave} = \frac{1}{3}(1.6 + 0.9 + 1.2) = 1.23$ ft

Volume of cut in triangle DEF: $V_{DEF} = 10{,}000 \times 1.23 = 12{,}300 \text{ ft}^3$

Total volume = 123,525 ft³ = 4,575 yd³           **A**

A longer (but more general) solution would subdivide the region into triangles, for example, ABC, ACF, CFD, and DEF.

## 621

Given that the side slope is 1:3, a vertical dimension of 12 ft corresponds to a horizontal dimension of 36 ft. Therefore, the plan dimensions of the top of the landfill are 3,928 ft × 2,428 ft. This gives a plan area = 9,537,184 ft².

The volume of the cover soil can be approximated as top area × depth = 9.5372 × 10⁶ × 1.5 = 1.431 × 10⁷ ft³.

*Note*: A more precise solution can be calculated by the formula of a pyramidal frustum, which shows that the approximate method yields only 0.5% error. This is because the thickness of the soil in question (18 in.) is so small compared to the plan dimensions.

Frustum of pyramid: $V = \dfrac{h}{3}(A_1 + A_2 + \sqrt{A_1 A_2}) \approx \dfrac{h}{2}(A_1 + A_2)$     **A**

## 622

Identify the edges of the limit of free haul by looking for two stations 300 ft apart which have equal cumulative yardage ordinates. Note that the mass diagram ordinates are equal (1,275 yd³) at stations 45 + 10 and 48 + 10.

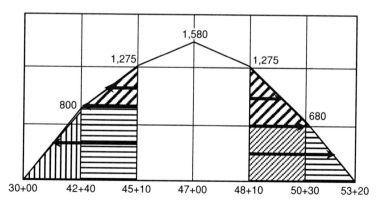

*(Continued on next page)*

The total overhaul is computed by taking moments of the mass diagram about the limits of free haul (stations 45 + 10 and 48 + 10) to the left and right, respectively, up to the balance points (stations 30 + 00 and 53 + 20, respectively). For each segment of the diagram, the overhaul is calculated as the product of the vertical ordinate difference ($\Delta y$) by the distance to the center of that interval. For example, the volume of 680 between stations 50 + 30 and 53 + 20 is triangular and the arm used extends from the limit of free haul (station 48 + 10) to the midpoint of the interval (50 + 30 to 53 + 20), which is at 51 + 75. These arms are shown with arrows in the figure above.

Overhaul = $475 \times 1.35 + 800 \times 8.90 + 595 \times 1.10 + 680 \times 3.65 = 10{,}898$ yd$^3$-sta

Cost of overhaul = $10{,}898 \times 3.7 = \$40{,}322$ **B**

# 623

522 working days at 6 days per week = 87 weeks

Number of weekdays = $5 \times 87 = 435$

Number of regular weekday hours = $435 \times 8 = 3{,}480$      Premium factor = 1.0

Number of extra weekday hours = $435 \times 2 = 870$      Premium factor = 1.5

Number of weekend hours = $87 \times 10 + 30 \times 10 = 1{,}170$      Premium factor = 1.8

Weighted average of premium factors: $\dfrac{3{,}480 \times 1.0 + 870 \times 1.5 + 1{,}170 \times 1.8}{3{,}480 + 870 + 1{,}170} = 1.248$

Labor premium = 25%      **A**

# 624

Material delivery = 950 yd$^3$/hr (bank measure), which is equivalent to $950 \times 0.85 = 807.5$ yd$^3$/hr (compacted) after shrinkage is accounted for.

Roller covers ground at 3 mph $\times$ 8 ft = 126,720 ft$^2$/hr.

As a 6-in.-thick (0.5 ft) layer gets compacted in four passes, each pass compacts the equivalent of 0.125 ft, which means it compacts $126{,}720 \times 0.125 = 15{,}840$ ft$^3$ (586.7 yd$^3$) of soil per pass. This is ideal rate of production.

Working at 80% efficiency, the roller compacts $0.8 \times 586.7 = 469.3$ yd³/hr

No. of rollers required = $807.5/469.3 = 1.72$. Therefore, 2 rollers are needed to handle the delivery of the material.                                                          B

## 625

From the curve, corresponding to a strength of 3,400 psi, we obtain TTF = 1,800°F-hr. Therefore, the time interval for sufficient maturity is given by:

$$\Delta t = \frac{TTF}{T - T_o} = \frac{1,800}{70 - 30} = 45 \text{ hr}$$

Contractor has to wait at least 45 hr after the concrete is placed before removing the forms.                                                                              C

## 626

From ACI-347, for wall forms, for rate of pour = 4 ft/hr and concrete temp = 60°F, base value (from tables, or equation) of the lateral pressure = 750 psf.

For type I cement with retarder, the adjustment factor = 1.2
For lightweight concrete ($\gamma = 135$ pcf), the adjustment factor $= 0.5\left(1 + \frac{135}{145}\right) = 0.966$

Therefore, the adjusted lateral pressure = $750 \times 1.2 \times 0.966 = 869.4$ psf

The hydrostatic pressure at the base of the form would be $135 \times 14 = 1,890$ psf. Therefore, the lateral pressure grows linearly from zero at the top surface to a value of 869.4 psf (at a depth of 6.44 ft) and then remains constant at that value for the lower 7.56 ft of the forms.                                                                          A

## 627

Forward Pass:

Activity A has Early Start Time (ES) = 0, Early Finish Time (EF) = 4

Activity D has only one predecessor (A). Therefore, $ES_D = 4$, $EF_D = 11$

Activity B: B has $ES_B = 0$, $EF_B = 3$

Activity E has only one predecessor (B). Therefore, $ES_E = 3$, $EF_E = 9$

(Continued on next page)

Activity G: $ES_G$ = larger of $EF_D$ = 11 and $EF_E$ = 9. Therefore, $ES_G$ = 11 weeks.

$EF_G$ = 11 + 3 = 14

Activity C: C has $ES_C$ = 0, $EF_C$ = 5

Activity F has only one predecessor (C). Therefore, $ES_F$ = 5, $EF_F$ = 11

Activity H has only one predecessor (F). Therefore, $ES_H$ = 11, $EF_H$ = 15

The project duration is the largest of early finish times for all terminal activities. Therefore, project duration is 15. G is also a terminal activity. Therefore, the late finish time for G is 15 weeks.

Therefore, total float for G = $LF_G$ − $EF_G$ = 15 − 14 = 1 week. **B**

## 628

According to the PERT model, mean and standard deviation are given by:

$$\mu = \frac{a + 4b + c}{6} \qquad \sigma = \frac{c - a}{6}$$

Complete the table with the mean ($\mu$) and standard deviation ($\sigma$) for tasks A, C, E, and F.

| Activity | Duration (weeks) | | |
|---|---|---|---|
| | $\mu$ | $\sigma$ | $\sigma^2$ |
| A | 4 | 0.33 | 0.11 |
| C | 6.83 | 0.50 | 0.25 |
| E | 6 | 0.33 | 0.11 |
| F | 8.83 | 0.50 | 0.25 |

Therefore, for the activity sequence ACEF, the mean is the sum of the means and variance is the sum of the variances: $\mu = 25.66$, $\sigma^2 = 0.72 \rightarrow \sigma = 0.85$.

For X = 25, the standard normal variable: $Z = \dfrac{X - \mu}{\sigma} = \dfrac{25 - 25.66}{0.85} = -0.776$

Probability of completion time being less than 25 weeks is calculated as:

$P(Z \leq -0.776) = 1 - P(Z \leq +0.776) = 1 - 0.7811 = 0.2189 = 22\%$ **D**

## 629

Current bid price can be calculated by applying inflation (3.2%) factor to price on file (2-½ years ago):

Cost per dump truck load = $525 \times (1+0.032)^{2.5} = 568$

Excavated material 323,000 ft³ will swell to $323,000 \times 1.25 = 403,750$ ft³ = 14,954 yd³. This will occupy $14,954/26 = 575$ truckloads.

Bid price should be $575 \times 568 = \$326,600$ for the 14,954 yd³ of earth moved.     C

## 630

Even though the problem is 3-dimensional, we can take advantage of the symmetry in the problem. All cables carry equal load. The length of each cable is given by:

$$L = \sqrt{5^2 + 8^2 + 2^2} = 9.644 \text{ ft}$$

The vertical component of each cable tension can therefore be written as:

$$T_y = T(8/9.644) = 0.83T$$

Therefore, the equation of vertical equilibrium is:

$$4T_y - 900 = 0 \Rightarrow 4 \times 0.83T = 900 \Rightarrow T = 271 \text{ lb.}$$     A

## 631

Design capacity = 40 tons

Since FS = 6, ultimate capacity = $40 \times 6 = 240$ tons = 480,000 lb

Pile hammer energy WH = 50,000 ft-lb = 600,000 in.-lb

$$Q_{ult} = \frac{WH}{S} + 1.0 \Rightarrow S = \frac{WH}{Q_{ult}} - 1 = \frac{600,000}{480,000} - 1 = 0.25$$

$S = 0.25$ in./blow $\Rightarrow N = 1/S = 1/0.25 = 4$ blows/in. = 48 blows/ft     D

## 632

The southwest corner of the mat is furthest (radially) from the wellpoint, at a distance:

$r = \sqrt{190^2 + 240^2} = 306.1$ ft.

Water table elevation at this point needs to be $325.65 - 5.0 = 320.65$ ft. Drawdown = $325.8 - 320.65 = 5.15$ ft.

At the southwest corner of the site, the radial distance is $r = \sqrt{240^2 + 300^2} = 384.2$ ft. Here, the drawdown is $325.8 - 324.0 = 1.8$ ft.

Applying the steady state equation to these two points:

$$Q = \frac{2\pi T(s_2 - s_1)}{\ln(r_2/r_1)} = \frac{2\pi \times 250 \times (5.15 - 1.8)}{\ln(384.2/306.1)} = 23,156 \text{ ft}^3/\text{hr}$$

$$= 6.43 \text{ ft}^3/\text{sec} = 2,887 \text{ gpm}$$

**D**

## 633

At a height $H = 14$ ft above the bottom of the footing, the width of the Rankine zone is

$$L_R = 14 \cot\left(45 + \frac{\phi}{2}\right) = 8.1 \text{ ft}$$

At a depth of 10 ft, the total overburden pressure = $120 \times 10 = 1,200$ psf

Active earth pressure coefficient: $K_a = \frac{1 - \sin\phi}{1 + \sin\phi} = 0.333$

Lateral pressure at a depth of 10 ft = $K_a \gamma H = 0.333 \times 120 \times 10 = 400$ psf

Lateral force on the blanket at depth of 10 ft = 3 ft $\times$ 400 ksf = 1,200 k/ft

Therefore, the effective length of reinforcing strip required (beyond the Rankine zone) is given by

$$L_E = \frac{1,200}{2 \times \tan 15 \times 1,200} = 1.87 \text{ ft}$$

Therefore, total length of reinforcing strip at depth D = 10 ft is L = 8.1 + 1.9 = 10 ft.

**B**

## 634

By employing one extra crew unit, work will be completed in another $90 - 15 = 75$ days

For these 75 days, an extra crew unit will cost an additional $= 3,200 \times 75 = \$240,000$

Early completion bonus $= 18,000 \times 15 = \$270,000$

Net bonus $= \$30,000$ (acceptable)

By employing two extra crew units, work will be completed in another $90 - 21 = 69$ days

For these 69 days, an extra crew unit will cost an additional $= 6,000 \times 69 = \$414,000$

Early completion bonus $= 18,000 \times 21 = \$378,000$

Net bonus $= -\$36,000$ (unacceptable)

By employing three extra crew units, work will be completed in another $90 - 31 = 59$ days

For these 59 days, an extra crew unit will cost an additional $= 9,000 \times 59 = \$531,000$

Early completion bonus $= 18,000 \times 31 = \$558,000$

Net bonus $= \$27,000$ (acceptable, but less than profit with one crew unit)     **A**

## 635

| | |
|---|---|
| Cement: | Weight $= 160$ lb |
| | Volume $= 160/(3.15 \times 62.4) = 0.814$ ft$^3$ |
| Sand (wet): | Weight $= 290$ lb |
| Sand (SSD): | Weight $= 100.7/105 \times 290 = 278.12$ lb |
| | Volume $= 278.12/(2.7 \times 62.4) = 1.651$ ft$^3$ |

Free water in sand $= 4.3/105 \times 290 = 11.88$ lb

| | |
|---|---|
| Coarse aggr. (wet): | Weight $= 420$ lb |
| Coarse aggr. (SSD): | Weight $= 100.5/103 \times 420 = 409.81$ lb |
| | Volume $= 409.81/(2.6 \times 62.4) = 2.526$ ft$^3$ |

Free water in coarse $= 2.5/103 \times 420 = 10.19$ lb

| | |
|---|---|
| Total water: | Weight $= 56 + 11.88 + 10.19 = 78.07$ lb |
| | Volume $= 78.07/62.4 = 1.251$ ft$^3$ |

*(Continued on next page)*

Total volume of these components $= 0.814 + 1.651 + 2.526 + 1.251 = 6.242$ ft³

This volume represents 97% of total volume (since air = 3%). So, total volume = $6.242/0.97 = 6.435$ ft³.

Unit weight = Total weight/total volume = 926 lb/6.435 ft³ = 143.9 lb/ft³      **C**

## 636

Forward pass through network results in: $ES_A = 0$, $EF_A = 3$, $ES_C = 3$, $ES_D = 6$

Free float of activity A = min ES of all successors (C & D) – EF of A = $3 - 3 = 0$      **A**

## 637

In the original plan (clear rectangles), E ends at 5 weeks 2 days, G begins at 5 weeks. Therefore, stage G can start without completion of stage E. Option **A** is incorrect.

Stage F is on schedule. Option **B** is incorrect.

Completion levels are (approx): A (75%), B (85%), C (85%), D (25%), E (25%), F (25%), and G (0%). At current time (5 weeks, 2 days), they should have been A (100%), B (90%), C (100%), D (75%), E (45%), F (25%), and G (0%). Stages A, B, C, and D are behind schedule. Option **C** is incorrect.      **D**

## 638

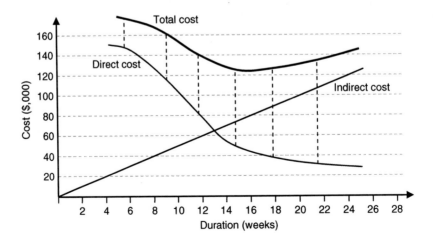

Superimposing the ordinates of the indirect cost (linear) on top of the direct cost curve, we get the total cost curve, which shows a minimum at about 16 weeks.

C

## 639

Slope of line defining boundary of tolerance = 1:500 (AISC Code of Standard Practice for Steel Buildings and Bridges)

Height at 4th floor = $4 \times 12 = 48$ ft = 576 in.

Lateral tolerance = 576/500 = 1.152 in.

D

## 640

Since the outriggers are simply resting on the soil (not anchored), the limiting condition is when the far side outrigger legs have zero reaction. For this condition, the inside legs (pair on the right) carry the total load (including crane weight = 50 k = 25 tons) of 65 tons. Taking moment about the inside leg, $40X = 25 \times 14$

Solving for $X$: maximum offset $X = 8.75$ ft.

B

## Answer Key for Construction Depth Exam

| | |
|---|---|
| 601 | C |
| 602 | B |
| 603 | C |
| 604 | D |
| 605 | C |
| 606 | C |
| 607 | D |
| 608 | B |
| 609 | A |
| 610 | B |

| | |
|---|---|
| 611 | C |
| 612 | A |
| 613 | B |
| 614 | D |
| 615 | D |
| 616 | D |
| 617 | B |
| 618 | A |
| 619 | C |
| 620 | A |

| | |
|---|---|
| 621 | A |
| 622 | B |
| 623 | A |
| 624 | B |
| 625 | C |
| 626 | A |
| 627 | B |
| 628 | D |
| 629 | C |
| 630 | A |

| | |
|---|---|
| 631 | D |
| 632 | D |
| 633 | B |
| 634 | A |
| 635 | C |
| 636 | A |
| 637 | D |
| 638 | C |
| 639 | D |
| 640 | B |

CPSIA information can be obtained
at www.ICGtesting.com
Printed in the USA
FFOW04n1531221015
17937FF

9 780071 777711